The Maker's Manual

Andrea Maietta and Paolo Aliverti

The Maker's Manual

by Andrea Maietta and Paolo Aliverti

Printed in Canada.

Published by Maker Media, Inc., 1160 Battery Street East, Suite 125, San Francisco, California 94111.

Maker Media books may be purchased for educational, business, or sales promotional use. Online editions are also available for most titles (*http://safaribooksonline.com*). For more information, contact our distributor's corporate/institutional sales department: 800-998-9938 or *corporate@oreilly.com*.

Editor: Patrick Di Justo
Production Editor: Kara Ebrahim
Copyeditor: Rachel Monaghan
Proofreader: Charles Roumeliotis
Translators: David Salvatori and Elisabetta Polcan
Indexer: Judy McConville
Interior Designer: David Futato
Cover Designer: Riley Wilkinson
Illustrator: Rebecca Demarest

April 2015: First Edition

Revision History for the First Edition

2015-04-03: First Release

See *http://oreilly.com/catalog/errata.csp?isbn=9781457185922* for release details.

978-1-457-18592-2

[TI]

Table of Contents

Preface .... **ix**

PART I. **The World of the Maker**

1. **Who Are the Makers?** **3**
   - The Culture of Reuse 3
   - We Are All Designers 4
   - Not Only Digital 4
   - The Maker 5

2. **The Origins of the Movement** **7**
   - The Culture of Sharing 7
   - The Triumph of Technology 8
   - The Fab Labs 8
   - The Spread in the Media 9

3. **A New Revolution?** **11**
   - The Introduction of Computers 11
   - The Power of Information 12
   - From Bits to Atoms 12
   - The Rebirth of the Economy 13

PART II. **Realizing an Idea**

4. **Can Creativity Be Learned?** **17**
   - Neurophysiology for the Uninitiated 17

The Learning Process 18
Techniques for Creativity 19
Lateral Thinking 19
Making Associations 19
Experimenting 19
Networking 19
Generating Alternatives 19
Changing the Assumptions 20
Shifting the Boundaries of the Problem 22
Pseudorandom Input 22

5. From Idea to Project 23
Design 23
The Design Process 24
The Problem Definition 25
The Requirements 27
Decomposition 28
Evaluating Alternatives 29
Aesthetics in Design 32
You've Got to Try, Try, and Try... 33
Eventually You Can Do It...Even Twice, Three Times 34

6. Project Management 37
What Is a Project? 37
The Project Manager 38
Management of a Project 39
The List of Activities 39
The Gantt Chart 41

7. Try, Fail, and Pick Yourself Up! 45
Business Plan 45
Abstract 45
Product (or Service) 45
Marketing Plan 46
Operating Plan 46
Management and Organization 46
Assets 46
Financial Plan 46
Ready for Success? 46
Wrong Assumptions 47
Success, This Time for Real 49

Customer Development ... 49
The Business Model Canvas ... 51

8. **Financing Your Work** ... **55**
Classic Funding Sources ... 55
The Friends and Family Network ... 55
The Bank ... 55
Alternative Solutions ... 56
Local and Regional Economic Development ... 56
The New Angels ... 56
Venture Capital ... 56
Crowdfunding ... 56
Beyond Financing ... 57
Bootstrapping ... 57
What's the Right Solution? ... 57

9. **Collaboration** ... **59**
The Importance of the Net ... 59
An Open Process ... 60
Distributed Intelligence ... 60
A New Protection ... 61
Bits, Bytes, and Atoms ... 62

PART III. **From Bits to Atoms**

10. **Managing Project Files** ... **65**
Distributed Design ... 65
Git and GitHub ... 66
Creating a New Project ... 66
The Three Areas of the System ... 68
Installing Git Locally ... 69
The Workflow ... 70
Not Only Trees Have Branches ... 73

11. **This Is Not a Pipe** ... **77**
Manufacturing Processes ... 77
Starting from Bits ... 78
Software ... 78
OpenSCAD ... 79
Hello, World! ... 80
Beyond Cubes ... 80

Variables 81
Move Slightly! 82
Lazy Is Good! 82
Other Transformations 84
Expanding OpenSCAD 87

12. **3D Printing** **89**
How Does It Work? 90
Materials 90
3D Printers 91
MakerBot 92
Kentstrapper 92
WASP 92
The Workflow 93
Corrections 93
Slice It Up! 95
Setting Up the Printer 95
Operating the Printer 96
What If You Don't Have a Printer? 98

13. **Milling** **99**
CNC Machines 99
Designing with a CNC 102
Software 103
CAD Software 103
CAM Software 103
Control Software 107
Where Do We Turn? 107

14. **Laser Cutting** **109**
Lasers 109
Laser Cutters 109
Models 112

PART IV. **Giving Life to Objects**

15. **Electronics and Fairy Dust** **123**
Hello World! 123
What You Need 123
A First Circuit 124
Current, Voltage, and Resistance 125

Circuits and Components 128
Circuits 128
Components 128
Creating a Circuit 135
Measurements 137
Ohm's Law 139

16. **Arduino** **141**
What Is Arduino? 141
The Software Structure 142
The Simplest Sketch 142
How to Upload a Sketch in Arduino 144
Interacting with the Physical World 145
Shall We Switch It On? 146
Not So Fast... 146
Pardon Me, You Were Saying? 146
Where Do You Store Your Data? 147
Only When I Say Go... 148
...Even Two, Three Times! 149
Beyond Digital 152
Some Exercises to Try 158

17. **Expanding Arduino** **159**
Reading the World: Sensors 159
Thermistors 159
Photoresistors 160
Other Kinds of Sensors 160
Changing the World: Actuators 161
Buzzers 161
Servos 162
Strong Currents 163
Shields 164
Smart Textiles 164

18. **Raspberry Pi** **167**
Component Check! 167
Getting Started 169
Basic Shell Commands 171
Operations on Files and Directories 172
Redirection 173
The World of the Superuser 174

Monitoring Hardware 174
The Graphical User Interface 175
Python 176
GPIO 177
Hello World 177
A Flashing Python! 179
Button, Button 181
Arduino and the Raspberry Pi 182

19. **Processing 185**
Your First Sketch 185
Let's Get a Move On! 188
How Many Circles? 190
I've Got the Power! 192
Programming with Cartoons 192
Classes and Objects 193
I Want...a New One! 194
OK, but What Should I Do with It? 195
Using a Drop 195
Raindrops Keep Fallin' on My Head... 196
Processing, Meet Arduino! 198
Libraries 198

20. **The Internet of Things 201**
Physical Computing 201
This New World 201
Where to Put the Data? 202
From Ivrea to Rome: Flyport 202
Raspberry Pi on the Net 203
Features of a Service 206

Index 207

Preface

A revolution is happening: the manufacture of objects is shifting from big companies—where your only choice might be the color black—to individuals, producing a previously unseen variety in things we make.

Today, thanks to versatile, powerful, and convenient tools such as Arduino and 3D printers, anyone can easily build, customize, fix, or improve objects. Tools and technologies have changed, but the passion for the process of creation hasn't.

A *maker* is not necessarily someone who works in design and manufacturing in her day job, but rather someone who always finds a way to turn her passion into an actual source of personal and economic rewards. She may get down to work to solve a personal challenge and then realize there are other people with the same problem. How can these garage inventors turn their passion first into a startup and then into a sustainable business, especially in this period of economic crisis? Even manufacturing and selling are changing—the old criteria do not work anymore—and people who have difficulties with change see their situation worsen day by day.

The new entrepreneurs have an utterly different approach: it is based on scientific techniques that were born within the industry field, reached the software world, and finally arrived in the business system. Just as the open source phenomenon influenced software in the 1990s, today open hardware and open design influence the production of physical objects; new startups create open source products, including both software and hardware. The enterprise philosophy itself is open. It is beneficial to cooperate and collaborate: people share ideas, and the more those ideas spread, the more the people and companies that are part of that community realize the profits. Anyone can contribute to projects and products and even create his own version, exchanging projects, ideas, and techniques to make (almost) anything. Such a model comes from the software world, where a worldwide community of developers works with a spirit of collaboration and sharing. Everyone benefits.

To become a maker there are many things to learn, many of which were familiar to our grandparents, although those skills have largely been forgotten now. A maker, like a modern Leonardo da Vinci, must apply a great variety of skills and knowledge, not only technical expertise.

This manual is an overview of the indispensable tools you'll need to become a maker: the starting point for a truly rewarding path.

Born from the actual experience of Frankenstein Garage (*http://www.frankensteingarage.it/*), which has been active with courses, workshops, and events for makers for years, this book explains complex concepts in a simple and intuitive way, anticipating the questions of those who wish to start or who still haven't managed to find their own way. The informal style helps you understand intimidating ideas, and takes you by the hand to help you create your personal toolbox, both physical and mental, in order to make the projects of your dreams a reality.

The manual consists of four parts:

- Part I discusses the makers, explaining the origins of the movement and its potential impact on the economic system.
- Part II proposes an easy yet structured approach to generate or perfect your own ideas (creative techniques, design processes) and make them grow in a favorable environment. It explains what a startup is, how to run a project, what innovation and business models are, how to find reliable collaborators, and how to raise financial resources.
- Part III is the more practical section, and briefly introduces the tools you'll need to collaborate. After that, you will learn how to physically create products starting from a model and using technologies such as milling, 3D printing, and laser cutting.
- Part IV explains how to give life to your creations, thanks to electronics and microcontrollers. We will also show you how to generate visual interactions, and will give you an overview of the Internet of Things (IoT), the new manufacturing frontier.

Have fun!

Book Site

We have created a website (*http://www.themakersmanual.com*) where you can find further information, resources, links, references, and anything else that we couldn't include in the book. You can also download all the sample code listed in the book from this website.

Conventions Used in This Book

The following typographical conventions are used in this book:

Italic
: Indicates new terms, URLs, email addresses, filenames, and file extensions.

`Constant width`
: Used for program listings, as well as within paragraphs to refer to program elements such as variable or function names, databases, data types, environment variables, statements, and keywords.

`Constant width bold`
: Shows commands or other text that should be typed literally by the user.

`Constant width italic`
: Shows text that should be replaced with user-supplied values or by values determined by context.

This element signifies a tip, suggestion, or general note.

This element indicates a warning or caution.

Using Code Examples

This book is here to help you get your job done. In general, you may use the code in this book in your programs and documentation. You do not need to contact us for permission unless you're reproducing a significant portion of the code. For example, writing a program that uses several chunks of code from this book does not require permission. Selling or distributing a CD-ROM of examples from Make: books does require permission. Answering a question by citing this book and quoting example code does not require permission. Incorporating a significant amount of example code from this book into your product's documentation does require permission.

If you feel your use of code examples falls outside fair use or the permission given here, feel free to contact us at *bookpermissions@makermedia.com*.

We appreciate, but do not require, attribution. An attribution usually includes the title, author, publisher, and ISBN. For example: "*The Maker's Manual* by Andrea Maietta and Paolo Aliverti (Maker Media). Copyright 2015 Edizioni FAG srl, 978-1-457-18592-2."

Safari® Books Online

Safari Books Online is an on-demand digital library that delivers expert content in both book and video form from the world's leading authors in technology and business.

Technology professionals, software developers, web designers, and business and creative professionals use Safari Books Online as their primary resource for research, problem solving, learning, and certification training.

Safari Books Online offers a range of plans and pricing for enterprise, government, education, and individuals.

Members have access to thousands of books, training videos, and prepublication manuscripts in one fully searchable database from publishers like Maker Media, O'Reilly Media, Prentice Hall Professional, Addison-Wesley Professional, Microsoft Press, Sams, Que, Peachpit Press, Focal Press, Cisco Press, John Wiley & Sons, Syngress, Morgan Kaufmann, IBM Redbooks, Packt, Adobe Press, FT Press, Apress, Manning, New Riders, McGraw-Hill, Jones & Bartlett, Course Technology, and hundreds more. For more information about Safari Books Online, please visit us online.

How to Contact Us

Please address comments and questions concerning this book to the publisher:

Make:
1160 Battery Street East, Suite 125
San Francisco, CA 94111
877-306-6253 (in the United States or Canada)
707-639-1355 (international or local)

Make: unites, inspires, informs, and entertains a growing community of resourceful people who undertake amazing projects in their backyards, basements, and garages. Make: celebrates your right to tweak, hack, and bend any technology to your will. The Make: audience continues to be a growing culture and community that believes in bettering ourselves, our environment, our educational system—our entire world. This is much more than an audience, it's a worldwide movement that Make: is leading—we call it the Maker Movement.

For more information about Make:, visit us online:

Make: magazine: *http://makezine.com/magazine/*
Maker Faire: *http://makerfaire.com*
Makezine.com: *http://makezine.com*
Maker Shed: *http://makershed.com/*

We have a web page for this book, where we list errata, examples, and any additional information. You can access this page at *http://bit.ly/makers-manual*.

To comment or ask technical questions about this book, send email to *bookquestions@oreilly.com*.

PART I

The World of the Maker

...people who hack hardware, business-models, and living arrangements to discover ways of staying alive and happy even when the economy is falling down the toilet.

—Cory Doctorow

- Chapter 1, Who Are the Makers?
- Chapter 2, The Origins of the Movement
- Chapter 3, A New Revolution?

Who Are the Makers?

1

Today we live in a world many of us define as "advanced," filled with technological wonders like smartphones and the World Wide Web. But these gadgets are just the fruit of an entire civilization based on the application of science and technology to our daily lives. Thanks to that civilization, we can live in a warm place, store our food without it spoiling, have light even when it is dark outside, communicate with the people we love anywhere in the world, or travel faster than our legs can carry us.

At the same time, many of these changes—the same ones that have improved our way of living—have limited our lives. Most people may say they cannot live without computers, telecommunications, electricity, and synthetic chemicals. If those technologies were to suddenly disappear, a large portion of the earth's seven billion people would start to die very quickly.

We are bombarded by media that do everything they can to encourage us to consume in an uncontrolled way—to queue in front of an Apple Store every six months, or to buy a new car every two years. And the same media make us feel "out of place" if we do not adjust to all the things advertising intends to inflict on us.

Within this context, products are no longer made to meet the consumers' needs, but to create a vicious circle: objects are designed to last shorter and shorter amounts of time, to break soon after their warranty expiration date (accurately calculated by statistics) so that we have to go out and buy new objects, thus artificially creating a market whose only aim is to support production.

Today governments are concerned only about GDP growth (in Italy, for example, the decreasing curve of the yield spread is, at the time of this writing, a further common concern). Even so, the GDP is a somewhat poor indicator of national contentment, because it also grows during events such as disasters or wars.

But has it always been like this?

The Culture of Reuse

For our grandparents and their parents, everything was different. Those born around, say, 1925, grew up during the Great Depression—a period of high unemployment, job insecurity, homelessness, and even starvation in some of the most advanced countries in the world. They learned themselves—and imbued their children with the spirit—to make do with what they had, which was almost nothing.

This shortage of resources led to a culture of recycling, respect, and reuse. Nothing was thrown away; instead, everything was ingeniously transformed using whatever tools were at hand. Our grandparents used to build what they needed themselves, and they were happy because they had something we often lack today: the sense of personal reward for having built something with their own hands, seeing their creation evolve from a conceptual idea to reality—from cutting boards, knives, barrels, and sickles to more technological tools (see Figure 1-1).

It was a question of culture: when something was needed or had to be solved, people tried multiple approaches, starting from what was available and often recycling it in previously inconceivable ways, until they found a solution. Then, as now, practice was the only way to actually learn.

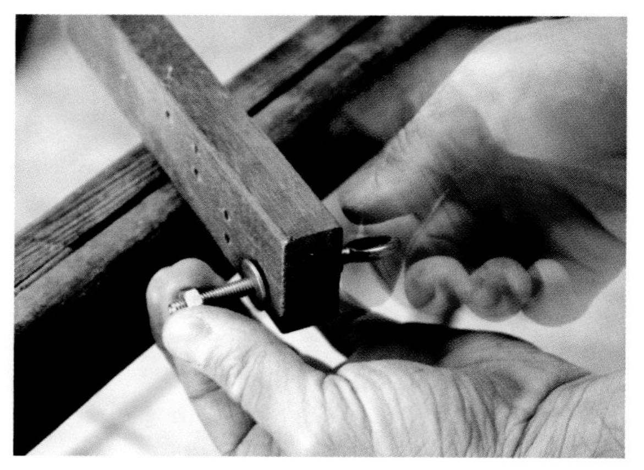

Figure 1-1 *The pleasure of building something with your own hands*

We Are All Designers

As children, many, if not all of us, dismantled some toys to understand how they worked. Some of us even managed to assemble them again. All the toys we dismantled taught us something, allowing us to modify them according to our tastes or to create new ones. In the past, this kind of activity was very widespread among adults too, practiced by so-called tinkerers: people who took abandoned objects and dismantled, modified, and redesigned them into something new and absolutely wonderful.

Today, technology allows us to do the same thing digitally. The necessary tools are at our disposal, free of charge or at reasonable costs. We can design very different objects following very similar processes. Thanks to our access to all information and to the community's support, learning is very simple, and we can become productive with different tools in a very short time.

Not Only Digital

In the 1990s everyone seemed to suddenly become a web designer: the spread of the Internet and the World Wide Web created a small factory of bits and bytes in many people's homes. With simple editing software, people could create websites. We believe that the immediacy of results and the low cost of entry have contributed to distancing today's young people, the so-called digital natives, from the traditional do-it-yourselfers who are still linked to the physical world.

What has changed recently is a sort of democratization in access to tools like 3D printers and other rapid prototyping machines, which has marked a return from bits (digital components) back to atoms (physical components) easier. These technologies have already been in existence for a long time, but they have usually been inaccessible to most people because of their extremely high costs. Today, a 3D printer can be had for as little as $500, much less than the original laser printer ($3,000). Even if other tools like laser *cutters* and computerized milling machines are still somewhat expensive, there are different services that allow you to use such tools at very low costs. It is like renting a factory without bearing all the startup costs: you only have to pay the manufacturing costs of what you need (plus, obviously, the supplier's mark-up).

This increased access to tools—as well as access to information on how to use the tools—has triggered the return to a culture of making and the spread of the maker movement.

The Maker

The maker is an enthusiastic hobbyist who gradually becomes part of a community of people who share the same interests. More and more he moves out of his field of competence, learning new skills thanks to the knowledge shared among the maker community. Once upon a time people had to apprentice with a carpenter if they wanted to create beautiful wood carvings, or with a blacksmith to forge metal. Today, those people can simply design objects with different shapes and have them created by computer-controlled woodworking machines or laser sintering machines.

Such hobbies are not only an occasion to meet new people, but they also might offer the makers the opportunity to earn some money or to found small companies, and in some cases they even lead to the birth of real phenomena in both cultural and economic terms.

Innovation—which, according to some economists, is the only way to increase a country's productivity—is a constant element for makers, as they always try to outdo themselves and go beyond what is at their disposal. The maker is like a new tinkerer, an inventor with a great deal of possibilities that, until recently, were inconceivable.

With this great power comes great social responsibility. We're fortunate that most makers tend to share the results of their work, and to collaborate with different people from all over the world, no matter their position or professional background.

Our grandparents were all makers. But what about us? Are we ready to be makers?

The Origins of the Movement

2

Humans have been makers since the dawn of our history. In fact, you could say human history began *because* we were such prodigious makers. Today, we are experiencing a renaissance in Do It Yourself (DIY) technology: the old maker tools of hammers, chisels, pliers, and tongs are being augmented by tablet computers, collaborative software, crowdsourcing, and desktop manufacturing. Sometimes the act of "making" is more digital, and all of these tools are replaced by a small portable computer.

The last 10 years have seen the growth of hackerspaces, makerspaces, and Fab Labs: workshops where lovers and creators of technology, mechanics, interaction, and art can meet, share their knowledge, and collaborate to create diverse objects (see Figure 2-1). In these places, it is possible to find—and use—equipment that is typically not available to individuals due to its high cost: drill presses, welding equipment, laser cutters, 3D printers, and more. With a reasonably priced gym-like subscription, anyone can access the equipment, which democratizes production.

In the beginning, the high initial cost needed to set up these spaces limited the expansion of this phenomenon, since only a few big institutions were able to finance this kind of workshop.

Figure 2-1 *A hackerspace plate in a picture by Vargson*

Today, however, there are thousands of such places. Even though they are typically found in universities and other institutions, commercial hackerspaces/makerspaces are growing. The most famous is TechShop (*http://techshop.ws*), which, as of this writing, has eight locations open in the United States.

The Culture of Sharing

The spread of digital technologies in the maker community and makerspaces has allowed the *early adopters* to be active in open source software projects, or at least to be familiar with them and share their philosophy. Sharing and collaborating are at the basis of the early communities that were taking shape within these spaces, which—thanks to the Internet—allowed them to expand and reach the remotest areas of the globe.

Many technologies that are adopted within these spaces can be dangerous if not used properly. Therefore, before accessing the equipment, beginners often attend training courses normally given by other enthusiasts. Training helps you understand a topic thoroughly, and is even important with activities without inherent danger, such as programming a microcontroller. Many makerspaces and hackerspaces have a culture based on a virtuous circle where mentors and students exchange roles, being a teacher of one topic and a student of another.

The Triumph of Technology

The easy access to digital technologies has fostered the spread of a new culture of making: sharing information—through the Internet—brings the manufacture of artifacts, even complex ones, within anybody's reach. Today we have the opportunity to turn our ideas into objects, transforming bits to atoms with a click of the mouse. We can access the power of a factory from our room, from a train, from the park.

The quick manufacturing offered by these new technologies reduces production time and cost, giving people with little experience and capital the opportunity to get quick feedback on various prototypes, thus fostering the incremental development process that is typical of a good project.

See, for example, the Gossamer Condor in "The Value of Quick Iteration" (*http://bit.ly/1wY4evP*).

The Fab Labs

In the late 1990s, Neil Gershenfeld, professor at the Massachusetts Institute of Technology (MIT), realized that his students were well prepared on theory, yet didn't know how to actually *make* objects. So in 1998, he created a course called "How to Make (Almost) Anything" (*http://fab.cba.mit.edu/classes/863.14/*).

In that course, Gershenfeld taught his students how to make small electronic circuits, how to program microcontrollers, and how to use Computer Numerical Control (CNC) milling machines, laser cutters, and other tools. The "almost" in the course title relates to, on one hand, the limits of the tools and materials, and on the other, to a number of shared values. Throughout the course, Gershenfeld realized that his students were using the equipment for their own purposes, rather than for their assigned projects. The creativity of those young students came as an extremely nice surprise: one student raised the curiosity of his fellow students by making a bicycle with traditional works and a frame made of laser-cut Plexiglas; another student, who used to feel discomfort when people invaded her personal space, created a smart dress that would raise spikes whenever someone got too close to her back.

Another student even created a cartoon-like soundproof backpack, in which she could scream and vent without anyone noticing, and then later release the recorded scream once she was out of the room.

From this experience, in 2002 the first Fab Lab was born. Fab Lab is short for "fabrication laboratory," a workshop where things are manufactured, but also for "fabulous laboratory." Gershenfeld has taken the Fab Lab culture around the world, helping local populations solve the issues of their communities: from the Norwegian shepherd who can locate his sheep on the mountains at the end of the grazing season thanks to a short-range radio transmission system, to the Indian farmers' village that doesn't have enough money to buy a tractor and makes do by adapting a motorbike, to the African farmer who pumps water out of a well by means of solar power. All these stories are collected in Gershenfeld's book, *FAB: The Coming*

Revolution on Your Desktop—from Personal Computers to Personal Fabrication (Basic Books).

The Spread in the Media

In 2005, O'Reilly Media published the first issue of MAKE, a magazine that today is the point of reference for the entire community of makers. Each issue includes articles and explanations, books and tools reviews, and, most of all, lots of projects, from the most basic to the most complex: a speaker in a cereal box, a rocket, a device that can throw the dog its ball when you're tired. The typical project can be carried out over a weekend, even though some can take much longer: for example, a makers' laboratory made from scratch has taken three issues. Most issues have their own theme: games, robotics, space, 3D printing, remote controls, and many others. Moreover, there are often articles for beginners that explain, step by step, the basics of different techniques as well as impossible challenges where, with a few (very few!) objects, the reader has to cope with the most absurd situations, nearly like being the ground crew on the *Apollo 13* mission, minus the pressure of actually being there.

One of the strong points of the magazine is its social aspect: many articles describe parent-child projects that can be easily carried out in a garage. Here, creating something together can bond a relationship that is crucial for the child's growth; other articles explain team projects.

To stress this social aspect even more, at the end of 2005, after publishing the first four issues of MAKE, on a late evening in the office, Dale Dougherty—one of the founders of the magazine—asked: "Wouldn't it be cool if we could get all these makers together in one place to share what they make?" It was a brilliant idea, and in 2006 the first Maker Faire took place in San Mateo. There, over 100 makers exhibited their creations. Since then, the Faire has grown every year (with over 1,100 makers and 130,000 attendees) and mini Maker Faires have popped up all over the world. In 2013, the first European Maker Faire landed in Rome, and worldwide there were 100 Maker Faires that year (see Figures 2-2 and 2-3 for photos from the 2013 Maker Faire Bay Area).

Figure 2-2 *Visitors at the Maker Faire Bay Area in 2013 (Alfredo Morresi)*

Figure 2-3 *There are all sort of things at the Maker Faire Bay Area (Alfredo Morresi)*

A New Revolution?

3

The first Industrial Revolution, which straddled the 18th and 19th centuries, was brought about by the introduction of machines into the production cycle—in particular the flying shuttle, which mechanized weaving, and steam power, which replaced human and animal muscle with a tireless engine.

This revolution changed everything. Living standards and education rose for millions of people. Global empires not only became possible but also, for the first time, practical. The earth's atmosphere saw its highest carbon dioxide levels in 800,000 years. Language changed, as words like *job* and *work* acquired their current-day meanings. Even the way we viewed time changed: before the Industrial Revolution the average person usually needed to know the time to an accuracy no finer than morning, afternoon, or evening. (Clerics needed to know the time a bit more accurately, to correctly perform the Liturgy of the Hours, the eight times a day a good Christian must pray; for that reason the best timepiece in a village was usually the clock on the church steeple.) But after work began being regulated by machines, people needed to know the exact hour—sometimes the minute—they had to arrive at work, the church clock was replaced by the factory whistle, and life became frantic.

The second Industrial Revolution dates back to the end of the 19th century—when electricity, oil, and chemicals led to the introduction of the assembly line in factories—and lasted well into the last third of the 20th century. Less well understood is the third Industrial Revolution, which we're in right now. It most likely began around the mid-20th century with the development of the computer, continued through the development of electronics, nuclear power, biotechnologies, nanotechnologies, and information technology, and it may very well be culminating with the development of desktop manufacturing machines.

The Introduction of Computers

When the Germans began using the electromechanical Enigma machine to encrypt their secret messages during World War II, the Allies needed a high-speed computation machine to decipher those messages. At first they used high-function mechanical adding machines operated by "computers," which were people, usually women, who punched the buttons. Unfortunately, this solution had some weaknesses: these computers, while proficient, were still performing calculations by hand, and, like all humans, needed time to eat, drink, and have a rest. Since the Allies were intercepting thousands of Enigma messages each day, this was not an easy strategy to pursue. So, the next step was the construction of the electromechanical calculator, which helped cryptographers to decipher the Enigma messages in less time. Toward the end of the war, the British developed the first all-electronic computer to decipher messages encrypted on a different German machine, the Lorenz SZ 40/42.

However, the first electronic computers had defects, too: they were large enough to fill entire rooms, extremely expensive, and very, very delicate, being made primarily of old-fashioned glass vacuum tubes. The Mean Time Between Failures (MTBF) for ENIAC, one of the first electronic computers (see Figure 3-1), was measured in hours in its early days. When it crashed, workers would walk *inside* the computer (which was room-sized, remember), and hunt for which of the machine's 17,468 vacuum tubes had burned out.

Figure 3-1 *ENIAC, one of the very first computers (US Army photo from Wikimedia Commons)*

Since these early times, computers have progressed in an impressive way: their calculation power has doubled almost every 18 months, so that today, any kind of smartphone is far more powerful than one of the first data processors.

The Power of Information

Great innovations often come from military projects, and the Internet is no exception. It started as a project called ARPANET, funded by DARPA (Defense Advanced Research Projects Agency, one of the agencies under the US Department of Defense) in the 1960s as a way to link the computers in four universities: Stanford Research Institute, University of California Santa Barbara, University of Utah, and University of California Los Angeles.

From that moment on, more and more computers have been connected with one another into networks: first locally, then at regional and national levels, and finally globally. We call this global network the Internet, the Net of the nets, which, thanks to its numerous services, has completely changed the management of information, and—with the development of the World Wide Web in the early 1990s—the way people and companies interact with one another.

It is easy to understand how the access to information remarkably enhances the power of all players involved. The consumer has a wider range of choices, the producer can deal with more markets and suppliers, and it is far easier to create new contacts and to collect information (via user feedback) regarding the quality of products and services. The computer and the Net can intervene at any point of the supply chain, improving it at all levels.

From Bits to Atoms

Today, thanks to the 3D printer, you can create a three-dimensional object just by downloading a ready-to-use file from sites such as Thingiverse (*http://www.thingiverse.com*) or YouMagine (*http://youmagine.com*) and printing it with a specialized device, just like we do with any paper document through a traditional printer. The concept is nothing new: the 3D printer was born long ago, but, until recently, the machines and materials were too expensive for an individual to own. Within the past few years, however, the cost of such devices has dropped so much that now it is possible to buy a desktop 3D printer (see Figure 3-2) practically at the same price as a laser printer.

Figure 3-2 *A personal 3D printer working*

In recent years, 3D printing and the maker phenomenon have captured the world's attention. They've had a strong impact on people's imagination: what would happen to the current economic system, some pundits wonder, if each of us could build a perfectly functioning object for any need? What would become of the system of industrial production and economies of scale if everyone could make the things they need? Some have even wondered what this means for the capitalistic system as a whole.

It is true that these changes in production processes as a whole will certainly influence the market and the global economy, but such an impact doesn't necessarily have to be bad. Certainly, targeted microproduction may help reduce the uncontrolled consumerism we have always been exposed to through the media, because it would enable us to start repairing things again just like our grandparents used to, instead of throwing them away. Distributed small-scale making will also lead to creating objects just where they are needed (for example, spare parts), instead of transporting goods all over the world.

Personal manufacturing must not necessarily be regarded as a threat to the economy and to industrial production, because it is often a component of them. The fact that it is possible to build something by yourself does not mean that everyone wants to do it.

On the other hand, many people are more likely to buy a ready-made object, available in the market, and then customize it according to their own needs, both in terms of function and/or of look. A quick look at models on Thingiverse or YouMagine turns up a lot of accessories for existing products.

The options for customization are endless—from turning a computer into a steampunk piece of art to engraving patterns on closet doors—yet these options do not hinder the mass-market production cycle at all. On the contrary, personal manufacturing can pave the way to a series of aftermarket services, thus allowing the makers—organized independently or in a network of new, specialized, digital technological craftspeople—to offer a customization service to those who can't or don't want to make the desired modifications personally.

The Rebirth of the Economy

The world is currently going through a serious period of crisis, especially here in Italy where, in 2013, youth unemployment exceeded 40%. In the face of this lack of jobs, a lot of people have started new companies, typically linked to web applications. Thanks to the Internet, the economic barrier to online entry has practically disappeared (even if the same can't be said about the Italian red tape and taxation system). Personal manufacturing may be a very important tool to help these young people—and not only them. Today, anyone may start their adventure as a maker and micro-entrepreneur from their home or garage, and the Internet can help them be global from the very beginning.

As it often happens, a maker may create an object for herself or another person and then realize that others might also want that object, perhaps with a few simple customizations. In such cases, personal manufacture is perfect because the setup costs of the machines are marginal, so the maker can carry out all those tiny modifications at a reasonable price. Customization of everything, from Ferraris to colored T-shirts bearing the image of a favorite singer, is a big business. What consumers are paying for is not just the cost of the modifications, but a special price for the opportunity to be different.

As soon as makers reach the point where a great many people are interested in their wares, personal manufacturing ceases to be sustainable and they must use industrial factories with mass production to exploit far more convenient economies of scale. A mold for injection molding can easily cost thousands of dollars, but at that scale the manufacture of a single item will not cost a few dollars (as it might with desktop manufacturing tools) but rather a few cents. This will enable the maker, collaborating with a contract manufacturer, to develop and grow without problems and without enormous startup investments.

Therefore, a maker following a sustainable business model can turn something conceived as a niche object into a mass-market product. Moreover, if a few of his products or services are successful, the maker can become an accomplished entrepreneur and create a solid company, thus bringing homemade competencies—with their shorter turnaround times and lower costs allowed by digital technologies—to more people's attention. Consequently, such a system can generate new jobs, help the local economy to restart from the bottom up, and maybe even cause a chain reaction in which more and more new companies are created and are able to generate, in turn, new jobs themselves.

PART II

Realizing an Idea

"Would you tell me, please, which way I ought to go from here?"

"That depends a good deal on where you want to get to."

—Lewis Carroll,
Alice's Adventures in Wonderland

- Chapter 4, Can Creativity Be Learned?
- Chapter 5, From Idea to Project
- Chapter 6, Project Management
- Chapter 7, Try, Fail, and Pick Yourself Up!
- Chapter 8, Financing Your Work
- Chapter 9, Collaboration

Can Creativity Be Learned?

4

Many of us have several ideas that we'd like to make real. Some of these ideas might have come to us after we worked on specific projects, or might be based on some research that we have carried out. Some ideas might have been a sudden spark inspired by an apple falling from a tree (hopefully not a MacBook Air!). Many people think that being an innovator and generating new ideas are natural gifts or talents: you either have it or you don't. The truth is, even if you think you're not creative, you can learn how to be.

Some people say that in order to innovate we don't need new ideas, we just need to stop thinking of the old ones. But where do "new" ideas come from? How can we be creative? Sitting under a tree, waiting for an apple to fall on our head, is of little use—even though, according to legend, we know about at least one distinguished precedent. To find an answer, we can turn to neurophysiology, the study of the functions of the brain and nervous system.

Neurophysiology for the Uninitiated

Just as asking someone "How do you feel about that?" doesn't make you a psychologist, reading this chapter on the brain will not make you a neurophysiologist. We're about to talk about an extremely complex field in an extremely simplified way. Nevertheless, it might help us understand our behavior a bit better.

Our brain is a wonderful machine, the most fascinating part of the entire human body. Sir Charles Sherrington, the "grandfather" of neurophysiology and a poet, used to say:

> It is as if the Milky Way entered upon some cosmic dance. Swiftly the head mass becomes an enchanted loom where millions of flashing shuttles weave a dissolving pattern, always a meaningful pattern though never an abiding one; a shifting harmony of subpatterns.

Our entire nervous system consists of nearly 100 billion neurons, the specialized, electrically excitable cells that process and transmit information. While there are many different types of neurons, they typically consist of three sections: a *soma*, or cell body; the soma's tentacle-like *dendrites*, which receive messages from other neurons; and a long-branched *axon*, which ultimately passes those signals to other neurons. While the signals within each neuron are electrical, communication between neurons is wireless, in the form of dozens of different chemicals that represent the messengers of thinking.

Every time a thought is born in our mind, thousands of these neurons trigger a very articulated sequence of actions and electric discharges: each neuron acts as a tiny yet extremely powerful data processing and transmission center that can manage a wide and complex flow of information. The path each thought takes through the brain creates a series of memory tracks—actual maps of our mind.

The Learning Process

The mechanisms behind learning have physiological roots: it's like the Colorado River carving out the Grand Canyon, where the rifts become deeper and deeper with the time passing (see Figure 4-1). They are shortcuts that make our life easier. The same kind of mechanism also comes into play in much simpler situations, making us do things without thinking. How many times have we found ourselves accidentally walking or driving to work or school on a day off, without realizing it? However, this is only a minor drawback when compared to the great advantage of using our autopilot to carry out everyday tasks by just following the mental models that we developed through experience.

This efficient system also has an unwanted side effect: innovating is *hard*. That is, it is hard to get out of the canyons that our thoughts have dug, because the resistance of known paths is much lower. Mental models are very useful, but they unfortunately limit our creativity.

An innovator, or what we call a "genius," has developed and learned throughout time a number of techniques that lead her to find new solutions.

Figure 4-1 *A river digs a deep canyon that is hard to get out of*

The continuous employment of these techniques—it sounds absurd, yet it is fully justified by the previously mentioned mechanisms—makes the innovation and creation of new models more and more natural. Many innovators apply these techniques unconsciously and, unfortunately, not everyone can describe them. Beginning chess players have to inspect the placement of the pieces on the board every time, to then try to evaluate all the possibilities. It is an immense task, so much so that even computers have a hard time with it. On the other hand, expert players have played so many games that they can assess positions and identify common or at least similar configurations very quickly, so they can decide, based on their experience, what is the best move or countermove. In chess, you need a lot of practice. But how can you reach high levels of innovation?

Luckily, somebody could explain many of these techniques, putting them at our disposal and allowing us to train our innovating skills. We can all be innovators, thanks to the processes and tools found, for instance, in the work of authors such as the great Italian artist and designer Bruno Munari, the mind mechanisms expert Edward de Bono, the art teacher Betty Edwards, and the essayist Tony Buzan.

Techniques for Creativity

We can "prepare the field" of our creativity by facilitating situations that plunge us into a specific topic, by surrounding ourselves with objects related to that topic, by stirring up our senses to prepare ourselves for the work of recombining and rearranging facts and ideas that we'll carry out on a subconscious level. Creative ideas or sudden flashes of intuition might come after days, weeks, or months of studying a certain topic. They might come in a moment of despair, when everything seems to go wrong. What is important is to plant the seeds of ideas, and to be ready when they finally ripen.

Lateral Thinking

Our culture values the sequential process of logical thinking, which starts with what we can see and proceeds step by step toward what seems to be the most natural solution. Psychologists call it *vertical thinking*: the thought process has a definite starting point and reaches a definite conclusion.

In 1967, Dr. Edward de Bono defined a different approach to problem solving, based on the investigation and research of alternative points of view. He called his approach *lateral thinking*. It starts with questioning the basic assumptions of a situation, diverting our thinking from the trails dug by our previous experience. Lateral thinking generates a high number of alternatives to explore, with the idea of moving on to vertical thinking only at a second stage, to develop the most promising ideas. You can be imaginative and train to develop this skill, but you may need some practice to get out of your usual schemes of thinking.

Making Associations

Steve Jobs, the incredibly creative cofounder of Apple, once said, "Creativity is just connecting things." If we really look at other people's innovations, we'll see that many of them are nothing more than a rearrangement of existing concepts and objects, often from seemingly unrelated fields.

For this reason, it is important to get out of our own competency zone, and look at the world from new points of view. We should draw from the diversity that new contexts can offer, to enrich our understanding of ideas and make the outcome better. It's a bit like playing music with friends and having the guitar player screaming enthusiastically at every track, "Think of it as funky! Think of it as funky!"

Experimenting

Exploring the world, visiting new places, trying out new activities, searching for information, and learning new things—all of these broaden our views. When we face a problem we have more alternatives, more choices that we can rearrange. Trying to make something with our hands, with or without help, is vital, because regardless of the outcome, touching and manipulating objects helps us develop a different, more physical intelligence.

Networking

Meeting new people who are out of our circle —people with different interests, backgrounds, and perspectives—allows us to exchange and share. It is also very useful to try to go for lunch, at least once a week, with somebody new or with some friend or colleague we don't see often: we might learn things we would never imagine. All the social aspects of the Internet can be very helpful, with tools like Twitter, Facebook, LinkedIn, and so on. Another useful activity is to attend conferences, exhibitions, and events dedicated to the themes that we are interested in or even just curious about; in these occasions, beside absorbing new concepts, we get the chance to exchange opinions with other people.

Generating Alternatives

Since childhood we have been taught to take notes in a linear way, using lists and recording facts, ideas, and situations to which we add our own ideas. As we have seen, having several alternatives to evaluate is one of the most efficient ways to find a good solution. Let's walk through a simple trick to generate alternatives. In the center of a sheet of paper, write a word that represents a problem and then draw branches around it, as shown in Figure 4-2. How many solutions would you like to find? Five? Well, draw five branches. Now, without thinking too much and in five minutes or less, try to write five possible solutions. Drawing the branches beforehand is crucial, because the brain tends to fill up empty spaces and complete images and sequences. To prove this, you could play (or sing) the musical notes from C to B and stop before completing the scale: don't you get a feeling of discomfort and incompleteness?

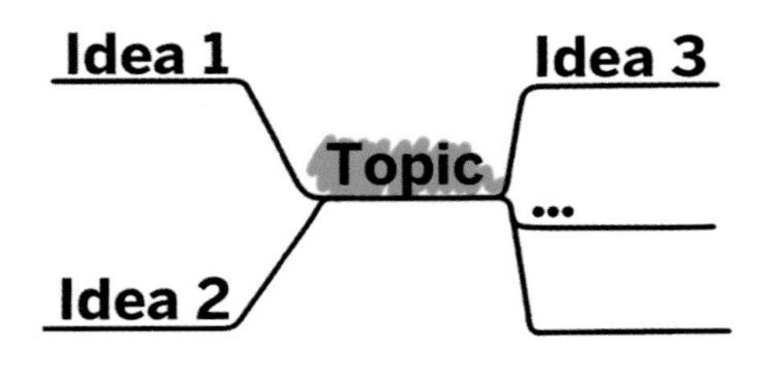

Figure 4-2 *We generate ideas starting from a concept*

You can shoot point-blank, without thinking too much and without feeling ashamed of thoughts that might seem silly or obvious. If, when the time is up, you haven't been able to find all the words you wanted, don't worry; just analyze what you *have* produced. This technique requires some practice; finding original ideas in a short time is not always easy to do immediately because we are chained to our thinking schemes.

This method, based on the nonlinear way we think, is the core of *mind maps*, a technique invented by Tony Buzan to capture thoughts in a more natural way, to take notes, collect ideas, generate new alternatives, organize activities, or make decisions (for more details, see "Mind Maps" on page 21).

Changing the Assumptions

Every problem starts with initial assumptions that we all too readily accept as facts. What would happen if we threw away those assumptions, and looked at the problem in a whole new way?

To do this, let's make a list of the features, qualities, and facts concerning our problem, as we see them. The simple act of writing takes us one step ahead of just having the information in our mind.

Once we have the list in our hands, let's try to debunk some of the assumptions and give the new composition some meaning. For example, here is a list about a bookstore:

A bookstore is a place where:

- We go to find a book
- There are many books
- There are tidy shelves
- There is a respectful silence
- You can buy books

Now let's try to overturn some of these statements:

- There aren't any books
- There isn't a respectful silence

What do we obtain? We might get the idea for a new kind of bookstore where books are not physically present and are printed on the spot, in a custom format.

Mind Maps

Mind maps are a model of how our thoughts are organized within our mind, with many branches and links between concepts and ideas. To make a mind map you need a sheet of paper, in the center of which you will write a word that summarizes the main concept you want to start from. Suppose you want to focus on sports, as in Figure 4-3.

Figure 4-3 *The first step in a mind map about a sport*

Now, draw some arcs and link the main concept to all the other ideas that come to mind, as in Figure 4-4.

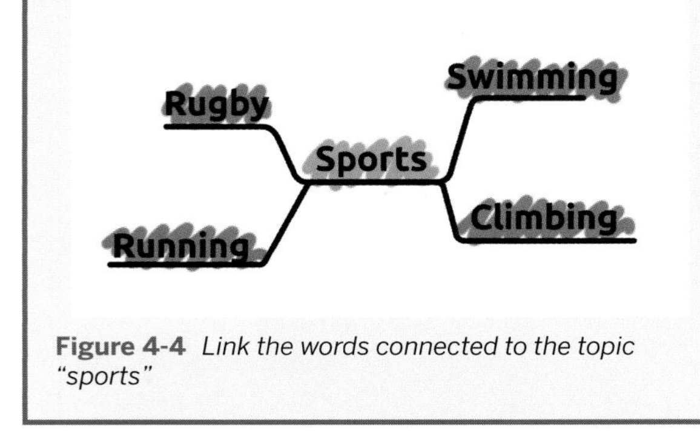

Figure 4-4 *Link the words connected to the topic "sports"*

Each concept awakens new ideas, not necessarily related to the main idea. You'll still include them, adding and specifying whatever crosses your mind, connecting the ideas in the right spots. Later you'll have time to thin out the map, if you wish, by eliminating those ideas that you do not consider useful for this investigation.

The mind map method is pretty simple. In addition to words, or even in place of them, you can draw pictures of things and concepts. You can color the map to enhance its visual impact, facilitating your ability to memorize and reason. With time and practice, you will draw better and better maps.

Often, if the map is very complex, it is worth creating several versions of it, each one better than the previous one, with a tidier structure. A further advantage of the iterative approach is that, by the time you go deeper into your explorations, you can see things in a clearer way, find new connections, and go in depth on those aspects that you consider most important.

Shifting the Boundaries of the Problem

As we have done with our assumptions, we can also question the boundaries of the problem: we can widen, shrink, and cross the "perimeter" we've arbitrarily set. What would we see on the other side if we leaned a little over the borders?

And what if we wiped them out completely?

Sometimes, in place of boundaries we have levels, doors, and lids. Learning how to accept anything as final and to question everything, even if it seems to make little sense, is a very important skill. In these moments we might feel somehow guilty, as if we were trying to "cheat." This is a natural reaction, because we are disrupting rules that we have gotten used to all our life—our mind objects, as if to warn us that we are breaking those rules. But many of those rules are self-imposed, often unconsciously, sometimes out of habit. Breaking them is one of the most powerful tools you have to see the world with new eyes and to find new solutions.

Pseudorandom Input

When we get stuck on a problem and it seems like there's no way we can find an alternative, it feels like we've fallen into a hole: it's hard to get out. In cases like these, the brain can benefit from a strong and unexpected input. Let's take the dictionary, open it randomly, point our finger at the page, and read the word we are pointing at. How can we link this word to our problem? What does this new, random input suggest to us?

The musician Brian Eno came up with card decks, called *Oblique Strategies*, to help his colleagues break out of their mental blocks. Before long, those and similar cards spread among other creative people, too. (There's some evidence that medieval thinkers used Tarot cards the same way—not to divine the future, but to use the random juxtaposition of images and symbols as ways to jump-start new thoughts.) There are even dice used to compose stories, called Rory's Story Cubes (*https://www.storycubes.com*) (see Figure 4-5), with objects on some faces and actions on others. All these methods can help us come up with new ideas.

Now you have no more excuses to say you have no imagination...you just have to train your brain!

So what's next?

Figure 4-5 *Rory's Story Cubes*

From Idea to Project

5

Once you've found the right idea, you have to give it a form to make it real. Its form includes both its obvious aspects and its hidden ones, such as architecture, mechanisms, and internal structures. For the design and implementation of such aspects, we can develop a process to guide our way. Even though some people seem to follow similar processes unconsciously, codifying it helps us understand what to do when we get stuck and don't know how to go on.

The path from idea to physical object is also typically constrained by available time and budget. You can make nearly any idea a reality, especially if you understand and can master process, time, and budget.

Design

In *Makers: The New Industrial Revolution* (Crown Business), Chris Anderson wrote, "We are all designers now. It's time to get good at it."

But what exactly is *design*?

We can give different definitions, but first we need to remember there are many kinds of design: product design, web design, graphic design, architecture design, software design, interaction design, industrial design, and so forth. The wide range of professional specializations and the different meanings of the word can cause some confusion. Is it possible to find a general definition to include *all* particular cases?

Karl Ulrich, professor at the University of Pennsylvania and designer of extraordinary experience, defines design as "conceiving and giving form to artifacts that solve problems," where *artifact* means anything manufactured by humans.

This definition derives from two others. The first one is from Edgar Kaufmann Jr., who was curator of the industrial design department at the Museum of Modern Art in New York and wrote that "design is conceiving and giving form to objects used in everyday life." The second definition is from professors Klaus Krippendorff and Reinhart Butter, who stated that "design is the conscious creation of forms to serve human needs."

So, the objective of design is to solve problems. People (well, hominids, anyhow) have had the ability to design and make tools for millions of years. Our desire to improve our living conditions and to avoid hard and repetitive work has led to solutions that improve our lot in life. This desire, and the ability to act on it, is the spur to progress.

We can see design as a component of the more general activity of *problem solving*: we all have problems to solve, but some of us do it as a profession, become *designers*, and specialize in a specific field. There are different reasons for this:

- On the one hand, it is practically impossible for someone to have all the competence and skill to design anything he needs.
- On the other hand, in many cases we are not merely solving just a problem of ours, but one shared by a wide category of users: we have to implement economies of scale in order to reduce development costs and consider the different needs within the group of users we are trying to help.

In this age of communication, we are accustomed to using tutorials, a sequence of steps composed of simple instructions to quickly learn something new. For design it is difficult to find tutorials, because it is a subject in which the different possible solutions must be explored and in which experience plays a crucial role.

For this reason it is challenging to find a magic formula or define a rigorous process that allows us to produce something beautiful and functional; nevertheless, we can try to define a process that remarkably improves our chances of success.

The Design Process

The design process, illustrated in Figure 5-1, consists of defining the problem we have identified, exploring the different alternatives for solving it, and choosing a plan to follow. This process is repeated until we find the best solution or until we run out of available time and budget (in which case, we'll have to make do with the partial solution we *have* reached).

The result of the design process is a plan to follow for the realization of the artifact that will solve the problem; imagine an interaction expert who explains how a website must be made but does not create it personally.

However, many designers—and practically all makers—do not limit themselves to the

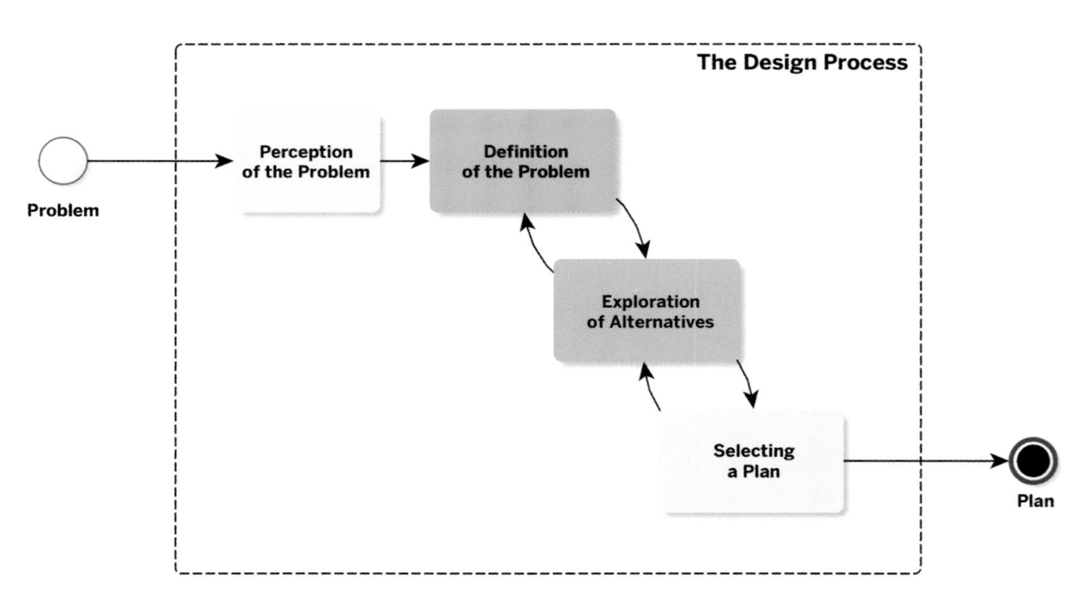

Figure 5-1 *The design process*

selection of a plan, but rather they also deal with the action-oriented final step, typical of the problem-solving process, yet sometimes alien to the world of design.

The Problem Definition

Once the existence of a problem (i.e., a *gap* in the user's experience) is identified, our first step is to try to define its nature precisely, since those who better describe the problem have higher chances of solving it.

Why?

Many people are so used to passively accepting anything that occurs that they hardly ever ask themselves *why* things happen. Maybe such a passive attitude lies in massive exposure to the television's cathode ray tube. But this is just a hypothesis and we may be wrong: in fact, in place of the cathode ray tube, today LED, LCD, or plasma screens are used. Asking yourself why things happen is at the basis of the *five whys method*, which together with other techniques forms Toyota's scientific approach to continuously improve the production process. It is a natural method that children often use:

Why is it hot? Because it's sunny.

Why is it sunny? ...

... and so forth

While children can go on like that forever and exhaust us, Toyota has decided to set a limit and move on to action at a certain point.

As its name suggests, the five whys method analyzes a problem, or, more generally, a situation, and asks *why* that situation is present, which progresses to a higher and more abstract level until the root of the problem is reached. Many assumptions have no solid foundation, but they are based on habit and laziness: "We do it this way because this is how we have always done it." Should we regard that as a valid and sufficient reason?

In the same way, we can ask ourselves *how* we can solve a specific problem. This process is very important to find the right level of abstraction and understand at what point we want or are able to intervene. For example, a frequent problem for many people who work on a computer while listening to music on their earphones is that they can't hear their office phone ringing. Imagine you want to create an artifact to solve the problem. At what level can you intervene? Let's start by understanding why it is a problem (if, indeed, it is):

- Why do I have to hear the phone ringing while I'm wearing my earphones? Because I don't want to miss calls.
- Why don't I want to miss calls? Because I want to be available when my colleagues call me.
- Why do I want to be available when my colleagues call me? Because I want to produce value for them.
- Why do I want to produce value for my colleagues? To be satisfied and improve the productivity of the company I work for.

We can stop here now, because it is already worth intervening and because we have reached an abstract level, even if we haven't gotten to the virtuous statement "we want to defeat world hunger" often proudly proclaimed by the best beauty pageant contestants! Now let's try the other way around:

How can I hear the phone ringing while I am wearing my earphones?

If we start from the condition that the earphones help us concentrate because they cut out any external noise and we don't want this to change, we are introducing a restriction. Considering this restriction, we can solve the problem by assembling a siren like the ones placed on fire trucks, thus scaring the life out of all our colleagues present in the office or even on our floor (see Figure 5-2).

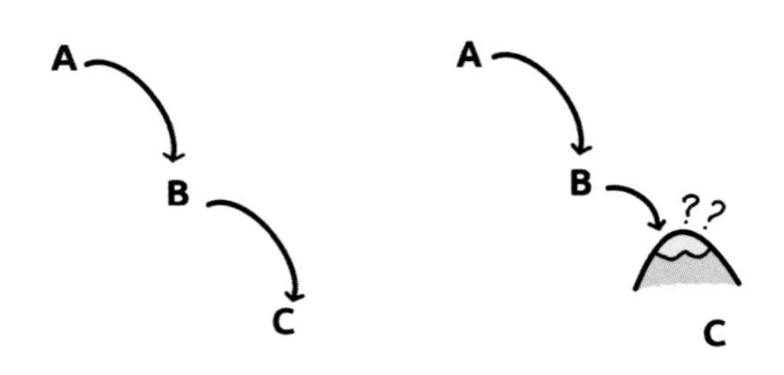

Figure 5-2 *It looks like we have reached a dead end...*

Or we take a step backward: the real problem is not the fact that we can't hear the phone, but that we miss our calls (see Figure 5-3). Let's now consider how we can solve it:

- How can I avoid missing calls? By noticing that my phone is ringing even if I can't hear it.

Now it gets more interesting...who said we have to hear the phone? We just need to know it is ringing:

- How can I notice that my phone is ringing even if I can't hear it? With a visual indication.
- How can I have a visual indication showing that my phone is ringing? By linking a light-producing circuit to the phone's cable.

And so on. At this stage we seem to be at a good point. If we had stopped when we identified the problem of hearing the phone, we would have never reached this type of solution, which may be useful for a different and wider market—for example, elderly people with hearing problems would happily avoid the surprise and hearing damage caused by a solution that merely raised the ringer volume to that of a fire siren. We have eliminated a *dominant idea*, a factor that paralyzed us and prevented us from reaching our objective. If we identify these factors as precisely as possible, we can avoid them and choose alternatives that will lead us to solve the real problem without wasting time building the wrong solution.

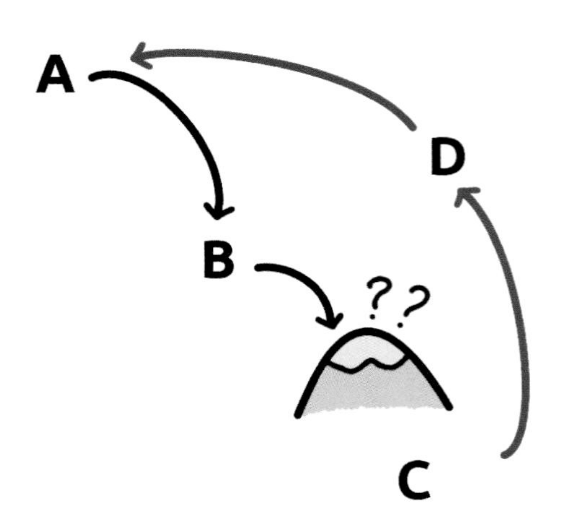

Figure 5-3 *Let's try to start from the bottom and find another way to reach the target*

Journalists use a similar method to that of the five whys. To be sure they have described a fact in a complete way, they check their answers to five, sometimes six, questions:

- Who?
- What?
- Where?
- When?
- Why?
- How?

Others make a very long list of questions (some as many as 50) and then try to answer or to use them as a stimulus to understand the problem better. As the famous economist Peter Drucker said in *The Practice of Management* (Harper & Row), "the important and difficult job is never to find the right answer, it is to find the right question." And the question, or better the *questions*, lead us to the requirements of the artifact.

The Requirements

After defining the problem, we can list the features we expect from an artifact; these are the characteristics that can help us close the gap (aka the problem we are trying to solve) in the user's experience. It sounds simple, but there are many aspects to consider:

- Different users have different requirements for the same problem.
- The requirements may be contradictory.
- Not all requirements have the same value.
- Not all requirements are of the same kind.

The first step consists of listing all requirements, which we will try to organize in categories later.

There are different methods to find out the requirements: there are some that we can follow by ourselves and others that involve other people, typically belonging to a segment we are interested in. In the example of the user wearing earphones while she works, what interests us are people who work at computers, obviously, but also elderly people, the hearing impaired, and maybe those families who have just welcomed a baby and do not want their ringing phone to disturb the newcomer's sleep.

One possible approach is represented by *focus groups*—meetings in which groups of six to eight people discuss the problem. They are much less effective than one-to-one direct conversations, because a series of sociological factors, not linked to the problem or its solution, get in the way. In both cases it is better to avoid closed-ended questions and favor open discussions in order to provide more and better data. It is not necessary to engage thousands of people—even 10 users for each segment of your user base are enough to discover most of the requirements. You as the maker need to be sure to do one thing: spend more time listening than speaking.

Direct observation is another excellent strategy, provided that the user is observed in his natural environment—that is, in the place and situation in which he perceives the problem, paying particular attention to nonverbal information. Observing people is simple, and we can do it in many ways: for example, in railway stations or supermarkets, where thousands of passersby can give us a lot of data to build a reliable pattern of people's behavior.

Observation must be carried out without prejudices and judgments; we observe how people move, interact, grasp objects, and cross rooms. Only later can we try to define a scheme; otherwise, our observations will be distorted and we risk taking a wrong turn in our process.

Observation is such a crucial tool that Scott Cook—founder of Intuit, a great American software house—said that "observation is the biggest game changer."

After collecting the data, you need to organize it in a series of requirements and follow some rules:

- Avoid excessive abstraction: requirements should have the same level of specificity as the data you collect.
- Express the requirements without implying a design concept, keeping yourself neutral toward the exact solution.
- Express the requirements as an attribute of the artifact that leads toward solving the problem.
- Avoid terms like *should*, *have to*, *could*, and similar because they imply a relative importance you have not evaluated yet.
- Organize the requirements hierarchically, picking up a main level that summarizes different secondary requirements.

Some of the requirements could be particularly important for the user, even if she doesn't realize it directly, but the importance clearly emerges from your research. It is better to highlight such *latent needs*—for example, by including an exclamation mark (!) before the requirement.

Let's consider the example of the computer user wearing earphones and try to list some requirements, organized hierarchically. If you have a name for an artifact, even if it's temporary, you can use it; otherwise, stick to the general term.

The artifact signals that the phone is ringing:

- The artifact ceases signaling when I answer the phone.
- The artifact works even if the ring volume is zero.
- The artifact can be seen working in the dark.
- The artifact can be seen working in full light.
- The artifact can be seen working even if we are not watching it directly.

The artifact is not expensive:

- The artifact is manufactured with an industrial production process.
- The artifact is manufactured with common materials.
- The artifact does not require unique working or assembling processes.

The artifact is safe in use:

- The artifact does not cause electromagnetic interference.
- The artifact does not cause electric shock to the user.
- The artifact is not dangerous for children.

...and so on. It is very easy to get up to a couple dozen requirements. In some cases you will even reach hundreds without a lot of effort. Yet, with all these things to consider, it gets difficult to choose a single one...so how can we make life easier?

Decomposition

Mental patterns tend to consolidate themselves, growing and merging with one another, and thus making it hard for us to explore all of the different requirements. We can try to divide the problem in many ways (see Figure 5-4). Some approaches are more natural, while others are less intuitive. The forced division obliges us to think about and organize the requirements in a different way. By decomposing everything into pieces we can even create new entry points, as we have already discussed with the five whys method.

Figure 5-4 *Possible decompositions for a problem, as well as possible decorations for a tile*

Sometimes we can divide our requirements in completely separate blocks, while at other times we will have blocks that can be superimposed and linked together in several ways.

Decomposition according to the user's needs

If, for example, we wished to design an ice cream scoop, we should enumerate a series of problems we can deal with *separately*. Let's choose some needs from the long list we have prepared:

- How to strain the wrist as little as possible?
- How to lift the ice cream easily?

- How to make the ice cream look nice in the bowl?
- How to manage to lift the ice cream even from the corners at the bottom of the tub?

With this approach, for example, we can evaluate the chances we have for not making the ice cream stick to the scoop without worrying too much about straining our wrist.

Decomposition according to a sequence of actions

We can decompose a problem according to the actions involved. If there is a presupposed *direction* or order, we can try to diverge from it. If we are designing a machine for picking up and wrapping vegetables, we can imagine an artifact that first wraps and then picks up vegetables. Or we may start the plants inside jars and then remove them. Does that sound crazy? In Japan they do this with watermelons, which they grow in cubic cages to save room and make them easier to transport (see Figure 5-5).

Figure 5-5 *Who says the best watermelons are oblong?*

Decomposition according to function

There are different ways of decomposing a problem according to function; to see this, we have to go back to the computer user wearing earphones. Let's try to concentrate on how to signal when the phone is ringing and on how to install the artifact near the computer monitor: we can evaluate several variables to make a list of possible solutions. Figure 5-6 shows some possible functional decompositions.

After evaluating the possible solutions to the subproblems you've identified, you can work toward a plan that will integrate what you've identified so far to create a prototype able to globally solve more problems (in an ideal world, all problems!). In this case, too, it is important to produce many options so that we can choose the most promising ones to use in subsequent stages. If you work on a team and you think of a brainstorming session, each member of the team should first explore the problem individually: statistically the team will produce a higher number of alternatives and at the end of the meeting the quality of the possible solutions will be better.

In the exploration process we use a representation, a pattern, a more or less formal language to describe the design we want to deepen. The representation requires a certain level of abstraction where we will describe the artifact, paying particular attention to the aspects that from time to time interest us more. In our work, we have used simple sketches, but you may want to use something more precise; for example, to design a new type of pacemaker, sketches wouldn't be enough.

But how can we evaluate the different solutions? Are there some criteria we can use in a coherent way?

Evaluating Alternatives

Design is not only graphics or decoration, but also—and especially—form and function. The result is a combination of function and form that represents a problem: while it is somehow possible to measure the functionality of an artifact, it is more difficult to find a formula or a criterion to evaluate its look and beauty.

Expert designers do not need to generate all of the possible solutions and combinations; just like an experienced chess player excludes a series of possible moves because he already knows that they will lead him to an unfavorable position, the designer "knows" which partial solutions to exclude beforehand in a given context and which combinations can work better than others. For example, although the *Terminator* (D) in Figure 5-7 has a very high "wow factor" that creates interest and amazement, it would hardly be chosen for further deeper analyses. Yet, if we asked designers to teach us their profession and explain the reasons for their choices, many of them would find it hard to answer because their knowledge derives from extensive experience and has been internalized into a kind of *tacit knowledge*. So we have to find a series of parameters that characterize the different possible solutions *and* a method to identify the most promising ones; this helps us select one or more plans to carry out.

Concept selection matrix

To help with your selection of concepts to explore, use a simple matrix in which you list a series of criteria with the corresponding evaluations for each proposed solution. Besides the criteria specific to the problem you're examining, the most typical evaluations are the *wow factor*, mentioned earlier, and *elegance*.

Your evaluations should follow as objective a method as possible, even if, because of the nature of the process itself, it is impossible to avoid any prejudice or personal inclination.

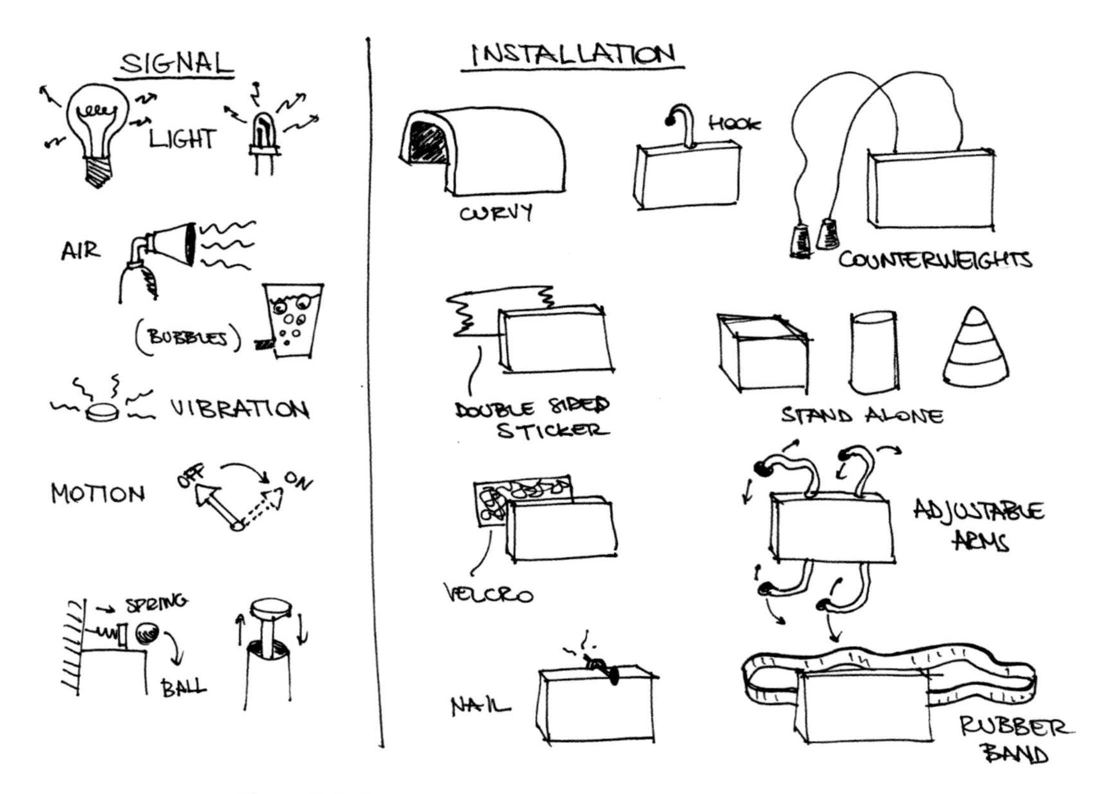

Figure 5-6 *Some possible solutions for a functional decomposition*

The matrix has at least three objectives:

- It helps you think in a structured way and solve the problem for a higher number of users.
- It communicates the logic underlying individual designs.
- It provides evidence of the reasons to prefer one design to another.

Let's go back to the phone example and observe how to build a partial matrix, paying particular attention only to the functions for which you have carried out the decomposition. At first we can choose a scale of three values: –1 if the evaluation is unsatisfactory, 1 if it is satisfactory, and 0 if it is neither.

In this case the most promising solutions from Figure 5-7 seem to be E and A, so we could exclude all the others, but it's good to consider as many as possible to establish contrast between alternatives. Table 5-1 shows a matrix of concepts.

For simplicity's sake we haven't considered half of the solutions hypothesized in the previous steps.

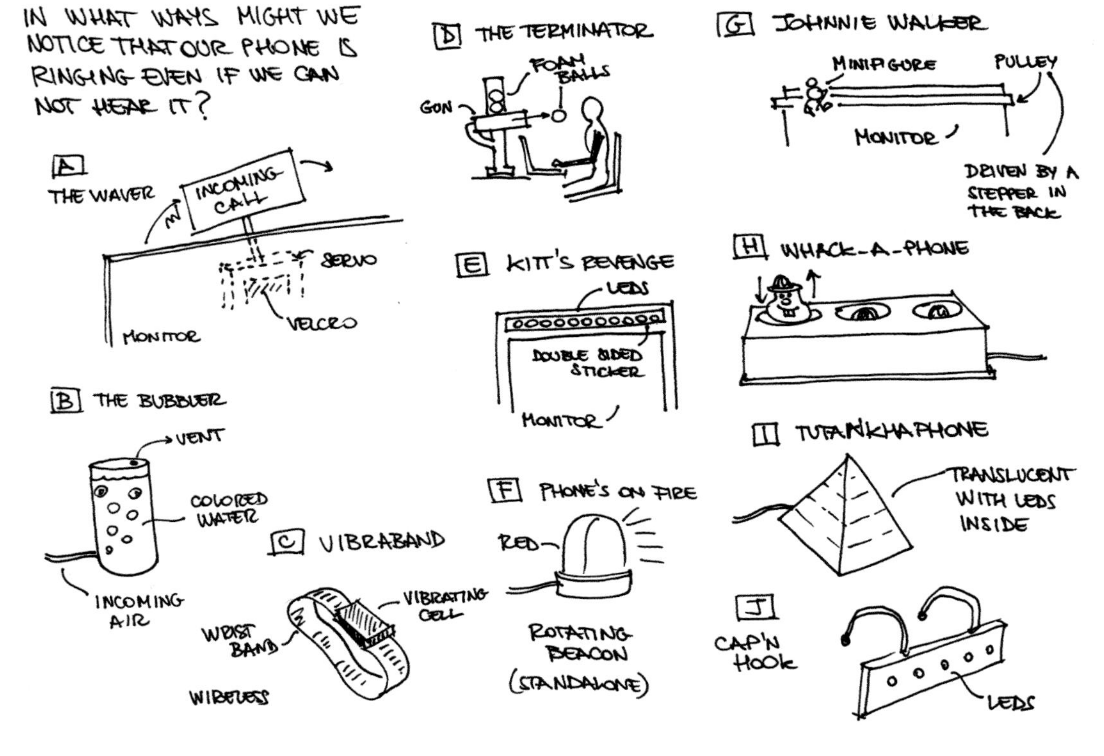

Figure 5-7 *Possible integrated solutions*

Table 5-1 *A simple matrix of concept selection*

Criteria/ concepts	A: The Waver	B: The Bubbler	C: Vibraband	D: The Terminator	E: Kitt's Revenge
Effectiveness	1	0	1	1	1
Ease of installation	1	1	1	–1	1
Minimal amount of space	1	1	1	–1	1
Cost	0	0	–1	–1	1
Wow factor	0	0	0	1	0
Elegance	0	0	0	0	1
Total score	3	2	2	–1	5

In some cases a three-value scale may not be sufficient, so you could instead use a five-value scale, from –2 to +2 or from 1 to 5. Whatever the case, it's important to use the same scale for all concepts. In addition, because not all requirements have the same importance, you could assign a *weighting* to each requirement. The simplest way to do so is to choose a percentage as weighting; so, the sum of all weightings must be 100%.

Aesthetics in Design

It is easy to understand why you'd want to include the cost as a criterion, but why add wow factor and elegance? Is it because you want to create, next to function and quality, an artifact with a beautiful look? The artifact's look may seem less important than its functionality, yet it is what users often pay most attention to without realizing it. The first impression comes from our perceptions, mainly visual, and, like it or not, it influences and conditions the whole decision process, even if the context, our experience, and our memory intervene at a subsequent stage. Even if functionality is important, the aesthetics of the artifact are fundamental.

The problem is that it is difficult to define what is universally beautiful, in particular according to a quantitative method. Moreover, aesthetics also implies a cultural component: just imagine the thousands of different ways in which art has idealized the feminine figure in different historical ages and cultures, in which different archetypes have emerged and continue to emerge. These currents are also linked to the evolution of ideas and values within a society, and they also constitute a psychological factor linked to symbols to which we assign values and qualities, which we in turn transfer onto the artifacts we have to evaluate. But how should we evaluate?

We can follow some guidelines inspired by biological factors. These tap into deep primal instincts, and although modern humans are much more rational than our ancestors, we still defer to millions of years of evolution:

- Objects whose shape reminds us of a snake acquire a negative meaning because they invoke the sensation of danger.
- On the contrary, smooth and shiny objects are perceived as positive because they remind us of bodies of clean water, a precious resource needed for survival.
- Finally, symmetrical objects invoke the concept of "healthy." It looks like Apple realized this a long time ago, and the company's sales prove it.

Those of us who have always been used to a technical approach may find the importance we place upon aesthetics strange and unconventional; however, it is not the technical part of the solution that wholly defines the success of an artifact. Sometimes we are tempted to use new technologies, new tools, and new styles just for the fun of it. When we do that, we are detaching ourselves from our real objective, just as when we think of an artifact only in terms of costs and manufacturing simplicity, forgetting its simplicity of use or how its functions are discerned (the artifact's *affordances*). We always have to place the users at the center of our process, because it is for them that we are trying to solve the problem.

If we are engineers in a highly structured process, that is OK: we can think of making controls, buttons, and displays in a convenient way, because (hopefully) someone else will deal with usability, aesthetics, and all the other aspects that are important for the user.

But if you're having to face the same problem as a maker or self-manufacturer, it is important to acquire an extended awareness and also consider all the aspects that go into making something.

You've Got to Try, Try, and Try...

At this point, before selecting a definite plan, we have to produce a *prototype*, an approximation of the artifact that comprises one or more dimensions of interest. Figure 5-8 shows a prototype of the "Waver" concept.

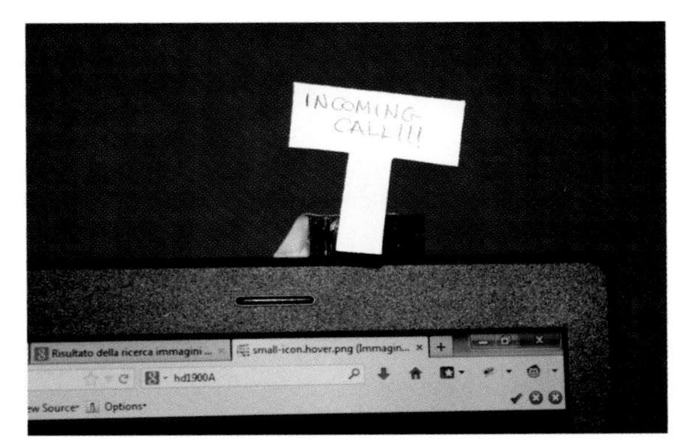

Figure 5-8 *A first prototype to inform us that the phone is ringing*

In this initial example, we have dedicated ourselves to understanding *how* we can catch the user's attention, neglecting the method that the artifact must employ to understand *when* the phone is ringing. Prototypes are very important and are used in each step of the design process. They have different objectives, but the most important ones are these:

- They allow us to answer questions such as "Will it work?" and "Will the user like it?"
- They communicate an idea.
- They are points of reference throughout the process.

In particular, prototypes enable us to understand whether we are following the right path. We may have misunderstood the user's needs or the user may have explained them incorrectly, or we may have not created an artifact able to completely solve a problem. Prototypes are crucial to take us back to the right path to reach our goal.

If we aren't sure how to proceed we can refer to *design patterns*: well-known and consolidated solutions that work reliably in a specific context.

Design Patterns

Architect Christopher Alexander is the first person to have identified and presented a *design pattern* in architecture—that is, a series of solutions to use in order to solve the most common problems or problems too difficult for beginners. By doing so, Alexander avoided having to reinvent the wheel with each project; for example, to create a room where we want to read and write, the best solution implies placing the desk next to a window.

There are patterns, collected in catalogues, for different sectors, even if the field in which they are best known is software development.

Each pattern is defined by four aspects:

- A name, crucial as a communication tool
- The problem, the context, and the application conditions
- An example of a tested and functioning solution
- The consequences of the pattern, which is extremely important because each solution is typically a compromise among the different forces in play

Patterns are not solutions; on the contrary, they are typically hints from which experts start to find a solution composed of a mix of different patterns. Patterns are to design as words are to language, and, as such, the more accurate they are, the more valuable the solution will be.

Eventually You Can Do It...Even Twice, Three Times

This entire process occurs in an iterative and incremental way: the more the design process develops, the closer we get to an optimal solution to the problem. The definition of the problem is clearer and clearer, we have learned to recognize the requirements, we have selected and evaluated some possible solutions, and we have made some prototypes. The more we proceed with the iterations, the better our artifact and the happier our user will be. The number of iterations required varies, even significantly, if we are facing a well-known problem (to build a new IKEA branch) compared to a new and unrepeatable project (such as Fallingwater, House over Waterfall, by Frank Lloyd Wright, shown in Figure 5-9).

Figure 5-9 *Fallingwater, House over Waterfall (public domain photo by Daderot, Wikimedia Commons)*

What is the difference between these two types of projects? Let's try to summarize it through a chart (Table 5-2) taken from the text *Agile and Iterative Development* (Addison-Wesley Professional) by Craig Larman, an expert in iterative and incremental software development. Of the two types of projects, which will require a higher number of iterations and the creation of prototypes with different levels of fidelity? It is quite easy to understand. In any case, the variety of opportunities and an iterative approach are always a great advantage for any kind of problem.

Once the best plan is selected, you can deal with the final creation of the artifact (see Figure 5-10).

You have different possibilities: start it immediately or plan it with care, do it by yourself or collaborate, save money on everything or always select the top of the range. What's the right way forward? And if the idea is not exclusively aimed at our use but at a whole market, how can we know if people will like it?

Figure 5-10 *A more advanced prototype to solve the problem of the computer user wearing earphones*

Table 5-2 *Predictable and innovative projects*

Predictable projects	**Innovative projects**
It is possible to collect all requirements and create the artifact accordingly.	Very rarely is it possible to provide, from the beginning, stable requirements in a detailed way.
At the very beginning of the project, it is already possible to estimate times and costs.	In the beginning it is not possible to make sensible estimates. As we proceed, empirical data emerge.
It is possible to identify, define, schedule, and order all activities in a detailed way.	In the beginning it is not possible to identify the activities needed. It is necessary to adjust according to what is being learned in iteration after iteration.
The change rate is relatively low and no unexpected event usually occurs.	The creative adjustment to unpredictable changes is common. The change rate is extremely high.

Project Management

5

Many of us have a secret idea that we would like to make real. Many makers have the dream of seeing their idea transform from simple entertainment or hobby into something more serious, which could initially be a small source of income and then, maybe, a full-time job. There are various reasons why you might make this decision: out of necessity, personal choice, because you can't find a rewarding job, to live in a more sustainable way, or to have more time to spend with family and friends. We'll call these kinds of big, transformative endeavors *life projects*.

It is true that a combination of strength, resolution, and intelligence can take us *closer* to our life project goals, yet these qualities alone are almost never sufficient to take us all the way. Knowing and using well-established time and project management techniques can help us avoid common pitfalls that await us on our life project journey. However, it is extremely important to apply these techniques to the right plan, because strictly following a useless plan won't get you anywhere.

So how will you get organized to realize your life project?

Completing a life project is like running a marathon. While there will always be moments of improvisation and quick decision making, you need to have a path marked out to take you from here to there; otherwise, you won't reach the finish line (or, even worse, you might get injured).

Turning the idea into reality is a matter of planning and organizing activities. There is plenty of literature on how to do this, but you can work out most of the information needed using logic and common sense.

What Is a Project?

We carry out projects every day, even though we may not look at it that way. Organizing parties, buying a car, writing a book for makers, whipping up lunch with the leftovers we have in the fridge—these are all activities with some common points that classify them as projects:

- Projects have a limited time duration: all projects have a limit, a deadline, or finishing time, from preparing dinner, which might require about 20 minutes or so, to the building of the Milan Cathedral, which took six centuries.
- Projects anticipate an outcome: an artifact, a product, or something nonmaterial like a service.
- Projects require the use of resources such as money and labor.

- Projects won't be repeated.

> *Even New York's Cathedral of St. John the Divine, under construction since 1892 and still not finished, will probably not be completed until the year 2200.*

The necessary activities to set up, say, a car factory with all the equipment and the assembly lines, constitute a project. The car production itself, on the contrary, is an activity that will be repeated over time and constitutes a process.

The resources available to carry out any project are limited. There are three key limitations: time, resources, and the scope of the project. These three aspects are strictly linked to one another and it is not possible to modify one without affecting the other two: for instance, if we drastically reduce the number of resources available to a project, such as people or materials, and if we don't intend to compromise on any of the requirements, the time to completion will necessarily increase. Conversely, if we decrease the time available to complete the project, or if we drastically expand the scope of the project, the number of resources we'll need will also change.

Since the late 1800s (and probably even earlier), project managers have been guided in their decision making by what was called the *iron triangle*, also known in recent years as the *Good-Fast-Cheap* triangle, shown in Figure 6-1. The three limitations of a project affect one another in a triangle of forces or influences. If you arrange the words *good*, *fast*, and *cheap* in a triangle, you'll see that to make a project good, it will be far from cheap and far from fast. To make it fast, it will be far from good and far from cheap. And if you want it cheap, it won't be good and it won't be fast.

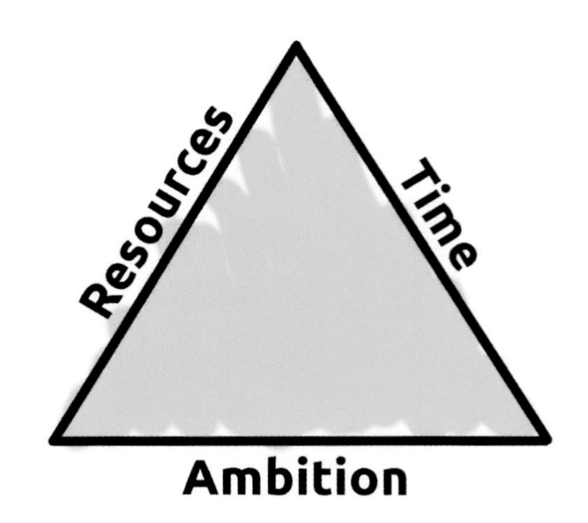

Figure 6-1 *The triangle of project resources*

The Project Manager

This all sounds complicated, right? That's why there are people who specialize in managing projects, called, unsurprisingly, project managers. Some project managers specialize in certain fields (aerospace, software, military, civil engineering, etc.), but some believe that a "real" project manager can successfully carry out any project in any field. In the software field, the Standish Group reported, in its 1998 CHAOS report based on studies of 23,000 projects, that a good project manager can account for 15% of the reason a project is successful.

Normally, the job of a project manager is not particularly creative: it requires preparing and writing a series of documents (among them the definition of the project scope), plans, and forecasts. It also calls for a constant overview of the work, in order to guarantee the necessary rigor and control that will lead to the final outcome.

At this point, a caveat is obligatory: in some contexts, such as the building of the Fallingwater house (*http://en.wikipedia.org/wiki/Fallingwater*), "being in control" is an illusion. What would happen if, halfway through a project, you realize that you are going down a completely wrong path? Would it make sense to go on and "successfully" carry out a totally faulty and useless plan? In some projects, the only constant element is change and our most effective weapon is our ability to adjust. Even the 34th president of the United States, Dwight "Ike" Eisenhower, used to say that plans are useless, but planning is everything.

Management of a Project

The lifecycle of a project can be divided into four stages:

Beginning activities
: This is the initial step in which you clarify and define the scope of the project, prepare an analysis, evaluate the work, and research the necessary resources.

Organizing
: Get ready to carry out the job, and define a precise plan with costs, timing, and outcomes. It's here that you acquire the resources and identify the risks.

Carrying out and monitoring the work
: The working teams are set up and they carry on the predicted activities. The project manager makes sure everything follows the plan. This is the most substantial step.

Shutting down the activities when the project is done
: The teams are dissolved, the product (or service) is delivered, and all commercial relations are ended. The best project managers carry out a retrospective evaluation of the whole process, the so-called *postmortem*, to sort out what they learned in the field and to consolidate their own experience.

While different schools of thought have different ways of managing a project, most agree that it is always necessary to perform an initial estimate and evaluation of the work. How can you do that, considering all the variables? You can use the ancient Roman strategy of *divide et impera*: divide and conquer. This breaks down the final product into many smaller units that are easier to understand (see Figure 6-2). You'll remember that we did the same thing during the design process. We really seem to have found a pattern!

The List of Activities

As with design, the act of dividing and subdividing the activities is iterative. We proceed by levels, going into more and more detail at each level. It is better to not specify clearly how the activities are going to be accomplished, but to restrict ourselves to merely listing the expected results. In this way, whoever is going to carry out the individual tasks will have more freedom, feel more involved, be personally and professionally rewarded, and, most of all, be able to reach unexpected results thanks to his creativity and innovation skills.

The hierarchic structure you obtain at the end of the division is called the work breakdown structure (WBS).

If you wanted to define a WBS to create the prototype of a radio alarm clock, start with the first level:

- Case
- Electronic board
- Firmware
- Electric certification

Then, go into details for each single point:

- Case:
  - Case design
  - Case printing
  - Assembly
  - Finishing and painting
- Electronic board:
  - Design
  - Board printing
  - Assembly

    — Testing
- Firmware:
    — Design
    — Writing
- Certification:
    — If needed: submission to the standardization bodies
    — Testing

Keep going until you reach microblocks:

- Electronic board:
    — Design:
        — Electronic calculations
        — Scheme design
        — Printed board design
    — Board printing:
        — Contact print
        — Board washing
    — Assembly:
        — Bending of components
        — Welding
    — Testing

The one problem with divide and conquer is that it's so easy to overdo. (If you're not careful, you may find yourself getting so detailed that you're specifying the number of screws you need for a particular task. That's going too far.) The entire purpose of this exercise is to make it easier to estimate the amount of effort the entire project needs, by estimating how much effort each activity needs. What is important is to follow the "100% rule": make sure to include

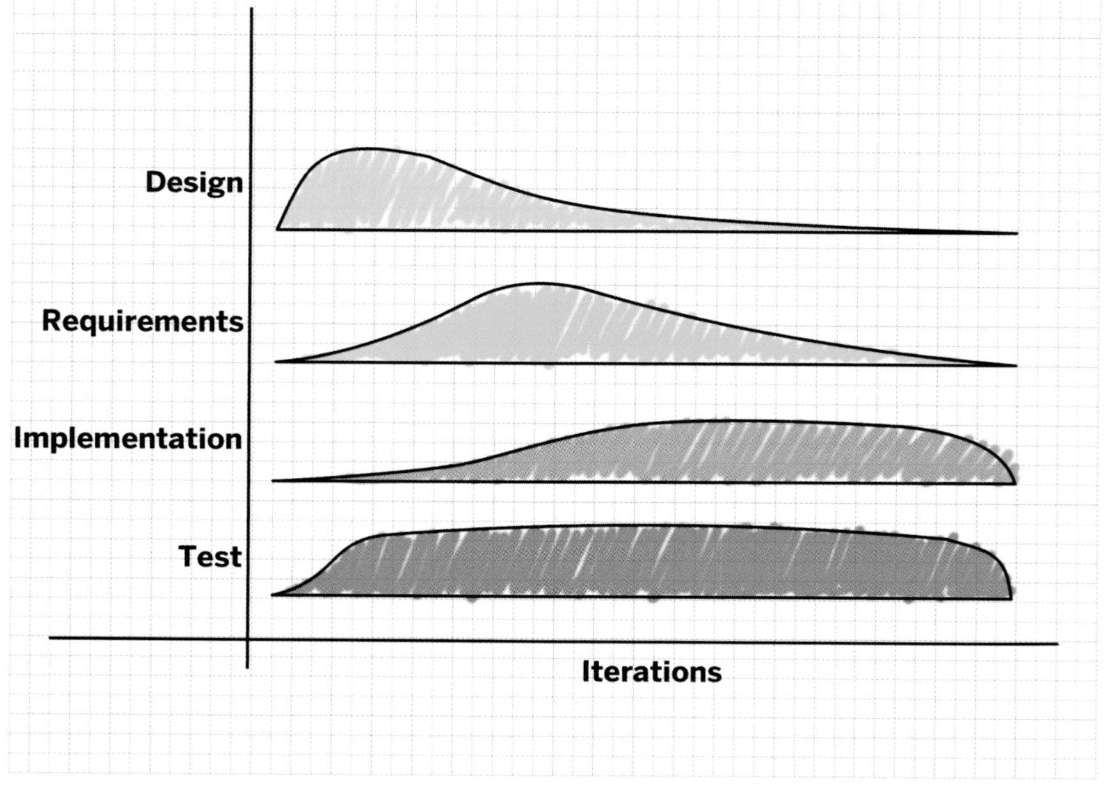

Figure 6-2 *Different activities in different moments*

all (and only) the activities that go into the project, both the ones you will do yourself and the ones you will ask someone else to do. It is important to always include the testing activities for the product or service.

WBS is just a starting point. You have to withstand the temptation of considering it a plan, because its elements have neither chronological order nor dependency ties. To aid in further planning, you need some more tools.

The Gantt Chart

At the beginning of the twentieth century, the American engineer Henry Laurence Gantt created a chart to support the management of his projects. He placed the time axis across the bottom, and the activity axis vertically on the left. Each activity was given its own row, and stretched horizontally from start time to end time. A project worked on by only one person looks like the steps of a staircase; the person completes activity A and moves down one row to begin activity B. When activity B is complete, he moves down one more row to activity C, and so on. In this way, all activities will proceed in chronological order, as shown in Figure 6-3.

If multiple people are working on the project, the chart becomes more complex. Certain activities can be worked on simultaneously, which should save time, but the project manager must devote more care to planning. If activity C needs input from activity A, they probably can't be done simultaneously, and the person assigned to C is idle until A is finished. Figure 6-4 shows the activities assigned to different people in different colors.

While assigning more people to a project usually means the project will be completed in less time, it also means the project is going to be more expensive. What is the best choice? There is no universal answer, because each choice is a compromise: in some cases, time is a critical factor so every expense is justified (think of the actions required to get people to safety after a natural disaster); other times, you can just wait. Moreover, you could give priority to some activities and you might want to carry them out before others, even if there is no link between them.

The sum of all the days needed to complete the project is called *effort*. Theoretically, if you get more people involved, the effort doesn't change (as a matter of fact, you need more coordination activities), but you can finish earlier. The time from the beginning to the end of the project is called the *elapsed* time.

The project manager checks on how the project progresses and updates her schedule

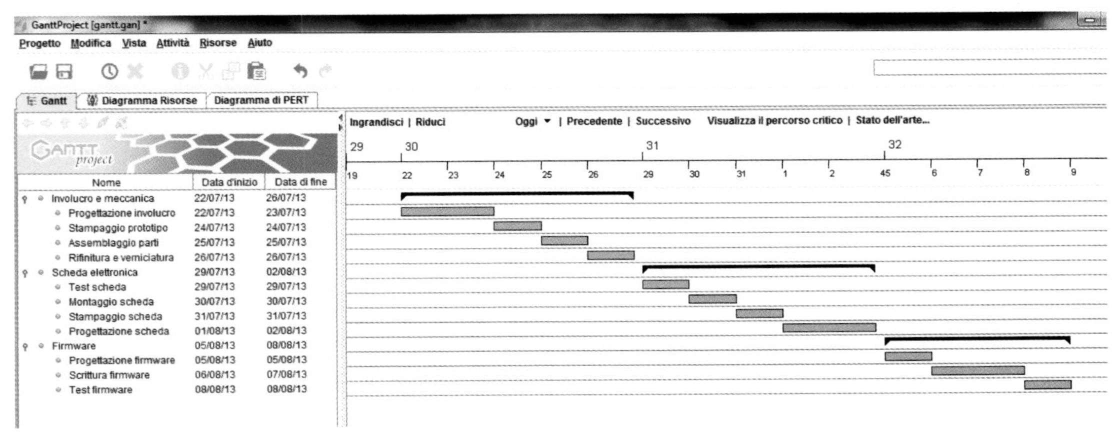

Figure 6-3 *A simple Gantt chart*

accordingly, highlighting diversions, if any, or delays and adjusting the plan to face any issues that arise.

Normally, there are several people who are interested in the project, with different degrees of involvement: they can be associates, partners, financiers, or clients. We commonly define these people by the general term *stakeholders*.

It is a good rule to periodically inform the stakeholders on how the project is developing, whether good or bad, so that they'll be able to make informed choices. These meetings are usually referred to as project review meetings (PRMs).

Throughout a project, you usually set some checkpoints called *milestones*, which, as their name suggests, mark especially important moments—for example, the presentation of a prototype to the client or the reaching of a midway goal.

According to Murphy's law, "if anything can go wrong, it will," so it is important to consider unforeseen events. There are a thousand reasons why the plan you set might change. A wise project manager identifies possible risks and mishaps that may occur and thinks ahead to the possible solutions.

This was extremely difficult to do back in Gantt's time, when everything was done by hand. We're fortunate that today you have a few software choices that can help. One standard is Microsoft Project, though there are many good alternatives, less powerful yet more than sufficient in the majority of cases. For example, GanttProject (*http://www.ganttproject.biz*) (desktop app) and Gantter (*http://www.gantter.com*) (available for free as a web application) are both able to manage projects created with Microsoft Project. You should also look at the open source Redmine (*http://www.redmine.org*) (web-based) and Planner (*https://wiki.gnome.org/action/show/Apps/Planner*) (GTK-based).

Even the best management software can't foresee everything that could go wrong. Thinking of the potential issues the project may face will make you ready if they actually arise. This kind of preparedness, the bible of which is the book *Waltzing with Bears* by Tom DeMarco and Timothy Lister (Dorset House), is called *risk management*. Demarco and Lister define this as "project management for grownups," because it goes beyond the naive vision of simply monitoring a series of activities.

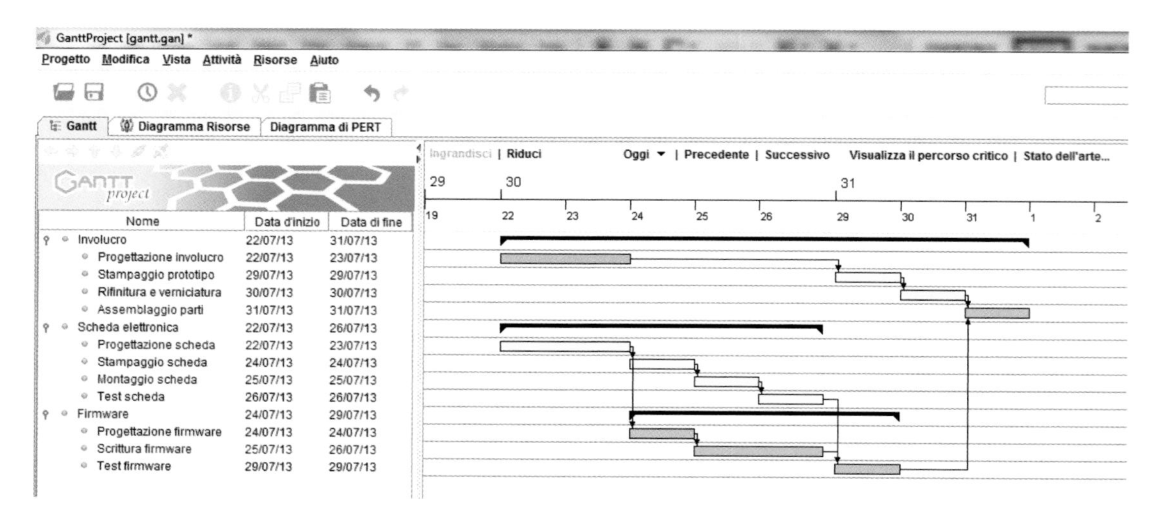

Figure 6-4 *A more complex Gantt chart*

One great aspect of being a maker is that we're always trying things we've never done before. This means that when we begin a project, we might not have any idea how long any of the subordinate activities will take, or what resources they will need. That's all right; the start of a project is an acceptable time not to know anything. Yet, because the introduction of a significant variation is expensive, it is better to assign priorities based on the respective risk of each activity, in order to limit the possible cost of an unanticipated change.

As you move on with the project, you learn new things and find answers to your questions, so the level of uncertainty related to the project decreases. The further you advance in the project, the more certain you become of its deadlines and costs. This is summarized in a chart called an *uncertainty cone* (Figure 6-5).

How do you want to carry out your plan?

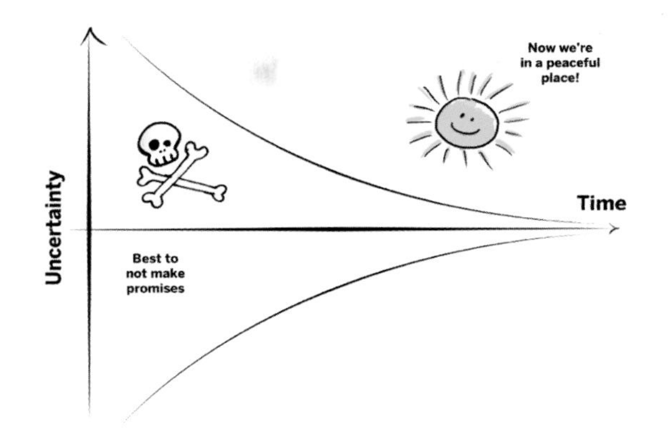

Figure 6-5 *The uncertainty cone of a project (Romina Paula Chamorro)*

7 Try, Fail, and Pick Yourself Up!

With a good idea in your hands, some well-written financial predictions, and the confidence that you can succeed, it is not difficult to convince someone to believe in you and finance your idea. In the blink of an eye, you can become an entrepreneur, abandon a boring and not-so-rewarding job in a cubicle, hand in your resignation—perhaps written a long time before—and leave, savoring your ex-colleagues' envy.

Business Plan

How will your new company work? What are your hopes and expectations? Is it worth it for others to invest in the project? Those who are thinking about financing you—that is, those who are thinking of becoming stakeholders in your adventure—want an answer to these and other questions, and they expect those answers to be presented as a document. This document is better known as a *business plan*. Let's summarize the main sections of a business plan.

Abstract

A business plan is a brief description of the product or the service you intend to create: basic concepts, costs and profits, some information on the market, and an analysis of the potential and originality of the idea. All business plans must begin with a concise and clear *abstract* (also known as an executive summary), without too many technical terms, because it is the first section to be read—or more likely skimmed—when outsiders are evaluating your plan.

Product (or Service)

In this section of the plan, you describe the product in a complete and detailed way, highlighting the innovative aspects and technologies used to create it, as well as the similarities to and differences from other products available on the market. When writing this part, try to avoid bragging along the lines of "We are so innovative that we won't have any competitors" because those who read the document are going to immediately think:

- You are not earnest. If a simple Google search shows other companies doing business similar to yours, the readers will assume you have not worked hard enough on your preparation.
- There is no market, so your idea is not interesting.

And remember, even if there is no competition now, a few months after the product becomes widely known, there'll be 10 other groups offering similar goods.

Marketing Plan

The marketing plan describes the market in which you'll operate, which can be an existing one (e.g., smartphone apps) or completely new (e.g., desktop 3D printers back in 2006). It describes direct and indirect competitors, actual or probable, in a more detailed way than the abstract does. To support your statements, you can also include analyses that present some market predictions. (Unfortunately, even the most meticulously done study cannot guarantee its predictions.) The marketing plan also explains who your customers are, and why they are interested in your product. It explains how you are going to present the product to the marketplace, detailing which channels you will use, and why.

Operating Plan

Now that you know what a Gantt chart is (see "The Gantt Chart" on page 41) it shouldn't be difficult to summarize and divide the whole project into macro-activities for the forseeable future, focusing on the main milestones. The plan must also be explained in words, expressly indicating priorities and limitations.

Management and Organization

Behind every project there is always a team. Who composes your team? What are the major positions in the business, and what role do they play? All those questions must be answered in this section. This is also where you would specify all your outsourced tasks: the company's lawyers, outside designers, contract workers, and the like.

It is also beneficial to indicate the company structure you intend to adopt: a sole proprietorship, an LLC (limited liability company), a limited partnership, a cooperative company, or some other type. The best strategy here is to engage a business consultant to get a clearer idea of the structure you want your business to have.

Assets

This section explains where the assets for the project will come from, and how you intend to finance yourself. The possible investors will be very interested in knowing how their funds will be used.

Financial Plan

At the end of the business plan you need a neat series of charts with financial predictions for the next three or five years, in which all the costs and earnings, as well as prices and sales hypotheses, must be indicated. You can easily estimate the costs by researching the typical costs of companies similar to the one you intend to start; for everything else, it's up to you to provide reasonably accurate figures. On the Internet you'll find many examples of financial plans, but if you are not an expert it is better to collaborate with someone who is good at numbers and can give you some advice. Sometimes you may be asked for a variation of the financial plan, called EBITDA, which stands for Earnings Before Interest, Taxes, Depreciation, and Amortization. In any case, you should "get your hands dirty" and understand what logic lies behind balance sheets and finance. You'll get back to actually making things soon, don't worry!

Have you got all you need? Are you ready to launch your startup?

Ready for Success?

You have come up with a sensational idea for a business, you've prepared a very detailed business plan, and you've convinced investors to give you startup funds. Now you are ready to start working! Find a prestigious building downtown (or an old abandoned factory, for that matter), buy the equipment you need, and recruit a team of experts to develop the product: engineers, designers, marketing experts, and salespeople. Don't forget the miniature golf and the ping-pong table—must-haves in the coolest offices!

The project managers make sure that the design and production departments work on the specifications you have provided them with, while the marketing department prepares its own plan, identifies possible customers, organizes focus groups, and prepares the documents on which your communications will be based. Meanwhile, sales managers and operators collaborate with the marketing department and prepare sales channels.

Everything goes on as it should, times and costs are respected, and the first prototypes arrive: the product is definitely good, so it'll be a success. Everyone is waiting impatiently for the launch date, which has been widely publicized by the media, thanks to your promotional campaigns and your PR staff.

At last the long-awaited moment has come: sales campaigns start...uh oh...it looks like in the first month you have achieved only 30% of what you predicted in your business plan. Well, that's normal; this product is really innovative, and it hasn't been understood yet. So you change your communication strategy and proceed.

After the second month, sales have dropped to 10% of your predictions, and in the following month you are practically at a standstill. Then goods are returned and sales decrease below zero; no new customers are to be seen.

In the meantime thousands of bills and invoices to pay arrive, suppliers complain, employees have the bad habit of demanding their salary, and investors lose their temper.

At this point, though, it is very difficult to get a second chance. Your glorious dreams have crashed against a smooth and extremely hard wall.

Why did this happen?

- Because you started work thinking only of your product
- Because you ignored who your customers are
- Because you don't know what your customers' problem is and how your product can solve it
- Because you have no idea of what is going to happen next week, let alone in three years' time
- *Because the whole approach we just outlined, despite being a well-trod path, is not suitable for a startup*

Wrong Assumptions

The process we have just examined is called the *product development model* and it is typical of already established companies operating in a well-known market. Unfortunately, even for these companies, the launch of a new product is often unsuccessful. But for them it is easier to get over the failure because the associated costs can be absorbed and covered by their other existing activities.

It happens in the best of families: just think of Sony MiniDisks, a CD-quality recording medium, and Sony Betamax, which had superior quality to VHS tapes. Neither of these technologies ever made a big splash with the public, despite their being released by a major company, and despite their technological superiority over their competition. For a startup, an embryonic company without a track record, the situation is even more delicate. A single bad event is sometimes more than the company can withstand. And yet almost all startups keep on following this model, with unrealistic expectations.

It is very common for startup business plans to contain product adoption curves (Figure 7-1) showing the company's customers in relation to time; these curves are described by Geoffrey A. Moore in his book *Crossing the Chasm* (HarperBusiness). The first customers that startups usually get are the enthusiasts or dreamers. Then come the mass of people who join in when "everyone else" does, and finally there are the pragmatic, late adopters.

Between the first customers and the mass there is the chasm, referenced in the title of Moore's book. This chasm is caused by the fact that the product—which is well suited for the first customers—must adjust to the expectations and tastes of the mass market, too.

Another serious problem is that too many Steve Jobs wannabes believe that they can do product development all by themselves, in closed rooms, guided by no more than their intuition, maybe with a little assistance from the folks in marketing. Since most of these people are not Steve Jobs, they fail because product development is not guided by the customer: the feedback from the customer, if any, comes only when the product is ready to be presented on the market. If any modification is required at this point of the project, it would be extremely expensive.

Many budding entrepreneurs believe in the motto "if you build it, they will come." This philosophy is extremely dangerous, because it leads you into a mental tunnel that you will exit only when you have finally created the product. While you're thus occupied, the world continues around you, possibly changing every assumption you had when you began the project. By the time the product is ready, the world has moved on and no longer needs your product (if it ever did).

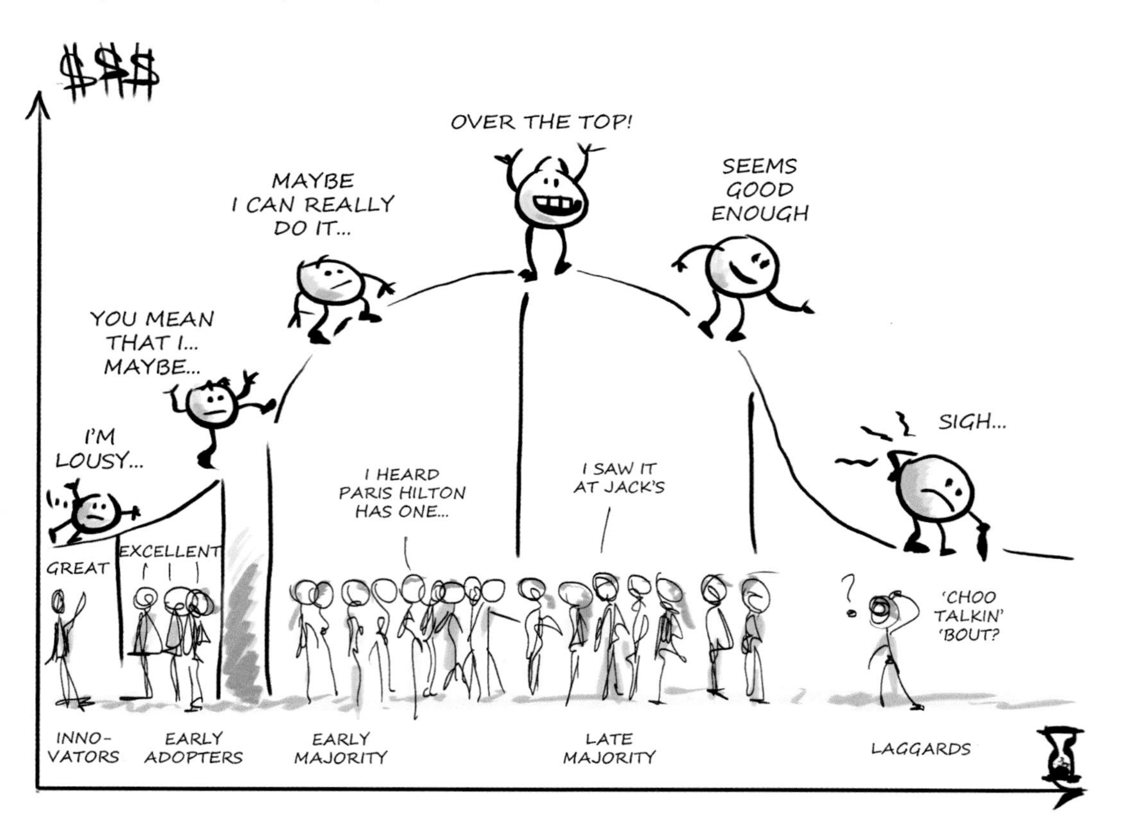

Figure 7-1 *The product adoption curve (inspired by a work of Tom Fishburne)*

Success, This Time for Real

Who are your customers? What are their problems?

You have to ask how you can help your potential customers to solve their problems. How can you organize yourself to do that? To avoid the most common mistakes, you must carry out the activities needed to find the answers to these questions well before starting product development, not at the same time. This research process is called *customer development*, and it was created by Steve Blank, a university professor with very broad experience in the startup field. Blank was awarded the title of Master of Innovation by the Harvard Business Review in 2012.

The starting point of customer development is represented by the hypotheses on which business is based. As we have seen, most of the time these hypotheses are nothing more than acts of faith. Building a business on faith means risking a lot.

A slightly more pondered and scientific approach includes the verification of the hypotheses before building a company on them. The most important assumptions, according to Eric Ries, entrepreneur and pioneer of the Lean Startup movement, are:

- The hypothesis of value: Has the product got a value for the customer?
- The hypothesis of growth: How does the adoption of the product occur? How can you discover and acquire new customers?

In order to verify the hypotheses, you have to conceive and carry out some tests. If you haven't got a product in your hands yet, you need to build one as soon as possible and with the minimum effort in terms of capital and energies. It need not be a finished product: a prototype suffices, which could be completely fake, or just a product with very few functioning characteristics, as long as they are useful for the team to understand what your customers need and to avoid creating a product no one is interested in. Eric Ries, in his book *The Lean Startup* (Crown Business), speaks of *minimum viable product* (MVP).

The aim of the tests is to learn; the *build-measure-learn* cycle (Figure 7-2) is a process that must be repeated continuously.

The figures you collect from the tests allow you to learn and adjust your plans, verifying and validating any decision on the basis of the actual consumer interest on the market.

That's the essence of the scientific method applied to business.

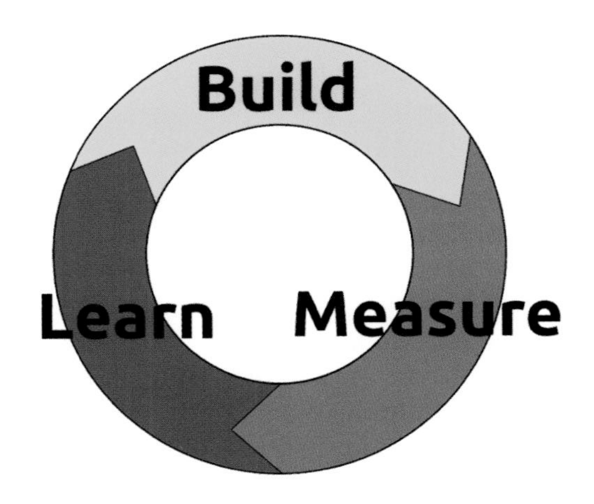

Figure 7-2 *The build-measure-learn cycle*

Customer Development

The process of customer development is divided in four stages: customer discovery, customer validation, customer creation, and company building.

It is the *customer* who is at the center of the process, and not your product.

Each stage of the process is represented by a circle with an arrow, as shown in Figure 7-3, indicating its iterative nature: repeat each step until you reach the desired result.

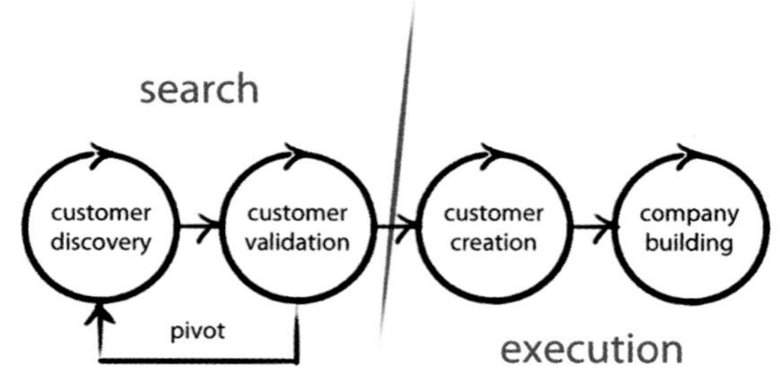

Figure 7-3 *The four steps of customer development*

Customer discovery

With this step, you try to understand who your customers are. It is not enough to try to put yourself in their shoes. Steve Blank says that we must get out of the office to observe and meet people; this idea is also present in the principle of *genchi gembutsu*, meaning "go and see with your own eyes," typical of Toyota's production system. First of all, you have to determine whether the problem you have perceived really exists and is felt by other people too: you can do this by speaking with people who you regard as potential customers, using all the methods you learned in Chapter 5. In particular, you have to listen, and avoid asking what they would like from the product or what they need, because in doing so you could divert them from their problem, which would lead them toward your preconceived solution. Instead, try to understand these potential customers, make hypotheses, and verify those hypotheses to ensure that your solution has value for them. Sometimes such a survey also makes you realize that your real-life customer is different from the one you imagined, which will lead to an evolution of the product and how you will develop it.

Customer validation

In this step, try to verify the business model you have hypothesized. The objective is to build a schedule that may be adopted by those who will deal with sales and marketing. Only when you have validated the different types of customers you're reaching for, and a sales process suitable for them, can you think of growing, progressing, and presenting the product to the larger public.

Periodically a pivot/persevere phase intervenes, during which you ask yourself if it makes sense to continue in the current direction or if a turn is necessary. Just as you make a pivot in basketball by keeping one foot still, when you make a business pivot it does not completely overturn the company; it simply modifies one or more of the fundamental hypotheses (e.g., your target market segment) and you go from there.

Pivoting is a difficult task: you risk throwing away months of hard work to start again. It may be useful to set some deadlines, or to react to specific metrics and gather everyone, with figures and data in your hands, to decide if it is indeed time for a turn. Waiting too long increases your risk of being off-track, while pivoting too often causes confusion and is not useful.

If customer validation fails, you must go back to the previous step, customer discovery. While you are learning it is normal to turn back, so you have to try to reduce waste as much as possible and always structure your tests so that they will teach you something and show you the direction to follow.

Customer creation

It is time to grow and create the final demand in order to reach the "masses" and guide your product to the sales channels. Usually, before passing to the mainstream customers, you will have to reexamine some concepts both in the product and in the business model, because these customers have characteristics and needs different from the early adopters you have dealt with so far. In this case too, it will be the tests that guide you.

At last it is possible to grow without problems, because you have already verified the product by considering the market and customers.

Company building

At this point your startup becomes "big" and restructures itself.

It's time to create formal departments like sales, marketing, and research and development. Now you have got it!

In this approach, you, the entrepreneur, must behave as a manager too. It may sound contradictory, as the two professions may seem poles apart: the manager is often seen as a rigorous controller, while the entrepreneur represents an eclectic and enlightened figure. In reality, a certain discipline and precision is needed to collect data, measure, and make calculations. Even if it appears hard and tiring (and it is), this is the price you pay in order to succeed.

The Business Model Canvas

Entrepreneur and innovator Alexander Osterwalder has conceived, together with Professor Yves Pigneur, what they call the business model canvas (Figures 7-4 through 7-6). It is a very effective tool to convey the essence of a business in a clear and immediate way. The canvas is divided in nine areas, one for each key point of a business:

- SGC = Customer segments
- PC = Key partners
- AC = Key actions
- VO = Value propositions
- RLC = Customer relationships
- RSC = Key resources
- CA = Channels
- STC = Cost structure
- FR = Revenue streams

The part concerning the customer segments, where you can include customers and companies you wish to reach with your product, is extremely important: if there were no customers there wouldn't be any business. They are a very important factor, and at first you know very little about them. You have to behave just like writers do with their characters: they know

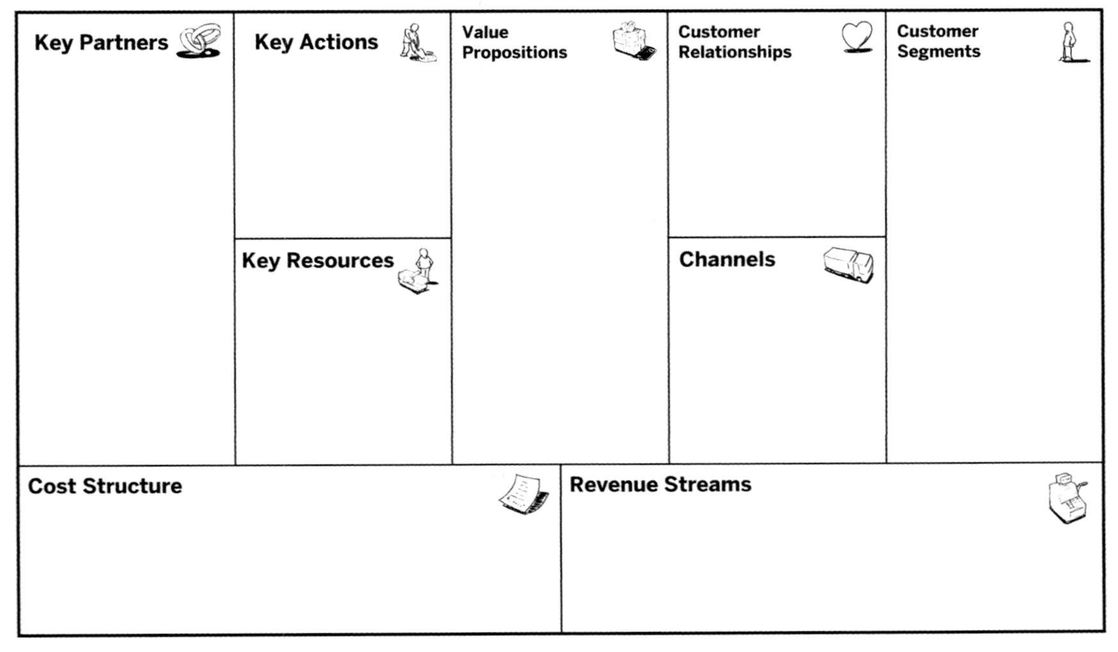

Figure 7-4 *The business model canvas*

them and they put themselves in their shoes so much that they can even guess what they keep in their fridge.

The value propositions section represents the value you are offering to your customers. The exchange of value occurs because the product is the answer to a problem or an existing need, and the customer recognizes in the product an intrinsic value higher than that indicated on the price tag. Many companies don't just sell a product or service: they sell an "experience" that their customers can live, which conveys value in a powerful and evocative way. For example, a car isn't sold as a simple hydrocarbon- or electric-based means of personal transport, but rather as a lifestyle statement.

The way in which a company communicates and reaches its customers is covered in the channels section. You can have your own channels, partners' channels, or a combination of these two possibilities; you have direct channels if the company has a sales force of its own, or indirect channels if the company uses third-party points of sale. A direct channel has the advantage of being controllable, compared to a channel with many mediators, but its implementation may be very expensive.

Once you have reached the customers, you have to establish and maintain customer relationships. You can choose a very personalized type of relationship or a completely automated one, managed with Internet sites and automatic exchanges, depending on the type of user experience you want to create. There are many possibilities; to get an idea, simply observe what other companies do. You will find sales assistants, personal assistants, account managers, self services, automatic services, users' communities, users/partners directly involved in the product development, and many other relationships.

If you have done everything properly, your customers will be willing to pay you. But how

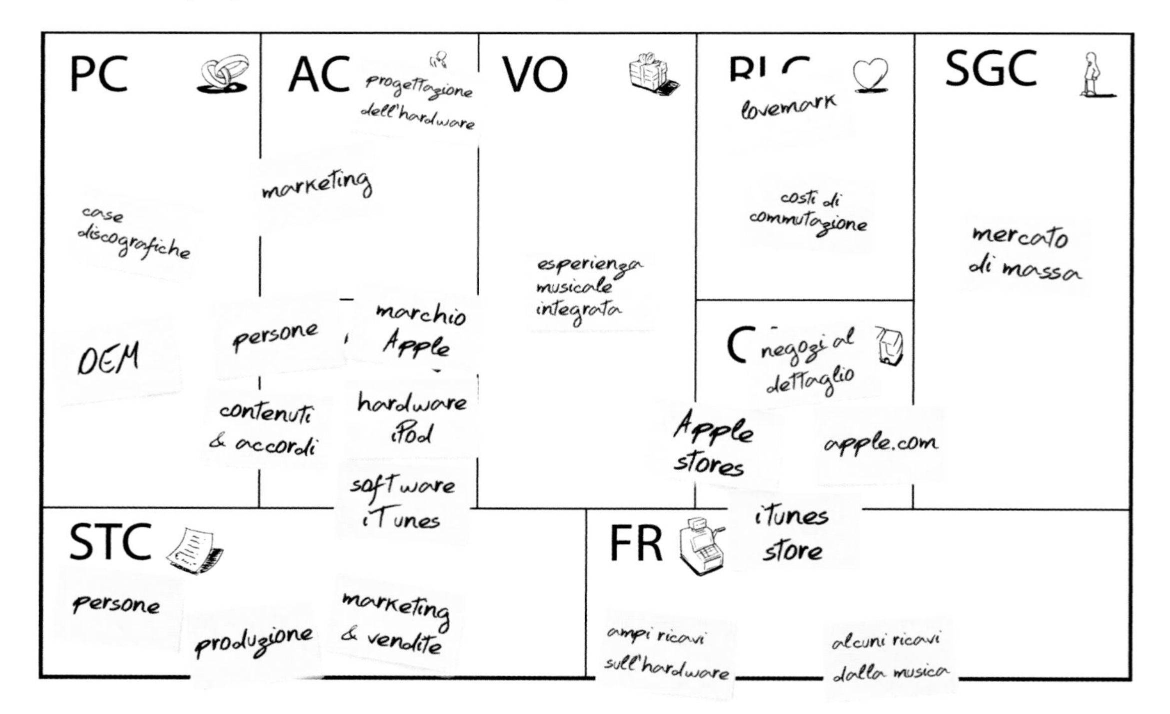

Figure 7-5 *A compiled canvas describing the business model of Apple iPod/iTunes*

much will they be willing to pay? And how will they do it?

You can think of different payment formulas and set prices in any number of ways. There are companies that offer some of their products for free, or sell them at a discounted price, and make their money from the sale of accessories (also known as "Give away the printer, sell the ink"). Others, like Google and broadcast television, give away most of their product for free and make money through advertising. Some others, especially on the luxury goods level, deliberately increase their price to attract the kinds of customers who place a lot of importance on image and brand.

To do what we have discussed so far, you need some key resources: the human, physical, material, intellectual, and financial resources that allow a company to create and distribute value. In addition, you will have to do some key activities: write software, create content, make machines work, and build objects.

In the wild world of business you are never alone; in the key partners' box, you will indicate the network of suppliers and collaborators that somehow help you make your business model work.

Resources, activities, and partners have associated costs, which can be found in the cost structure section of the canvas. To study this topic more in depth, we recommend the book *Business Model Generation* by Alexander Osterwalder and Yves Pigneur (Wiley).

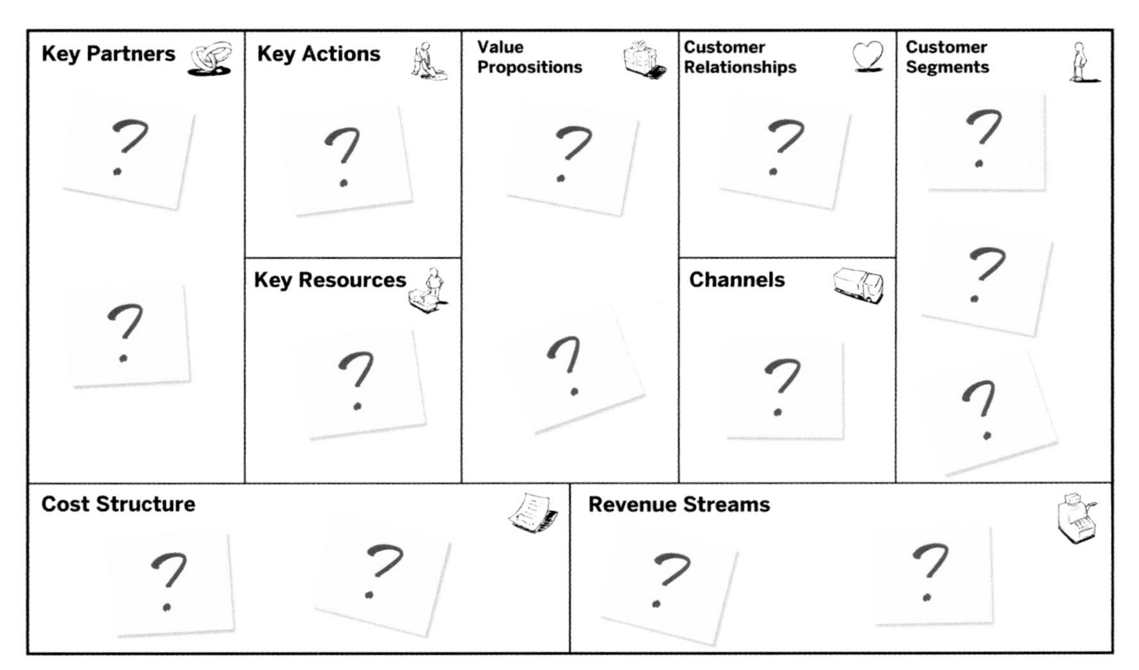

Figure 7-6 *Until we validate them, our boxes remain mere hypotheses*

Financing Your Work

8

In Chapter 7, we said that an enterprise needs key resources, activities, and collaborations. This all comes at a cost and, if the maker wants to turn into an entrepreneur, she must understand how to face that.

Classic Funding Sources

Two traditional financing sources are:

- Asking friends and relatives
- Asking a bank

Let's look in detail at these.

The Friends and Family Network

Many small entrepreneurs start by investing their savings and, for any remaining needed funds, asking parents, relatives, and lifelong friends.

It is often fairly easy to convince them: they are people who have seen us grow up, they love us, and they want us to be successful. They are already set to believe in what we believe in, they get carried away by our enthusiasm, and they are going to share our vision. Typically, they won't ask for interest, and in some cases they won't even want their money back.

If our startup turns into a proper sustainable business, that is great. If things go bad, chances are we'll drag down the people who believed in us. In these cases, the issue goes beyond the economic factor: even the most strongly established relationship can become strained when money is lost, leaving the entrepreneur alone in a hard moment, without any moral support.

The Bank

Another solution is to go to a bank and ask for a loan. In nearly every case the bank will ask for some form of collateral to secure the loan, for which you might end up asking your family for help anyway. In this case, the list of people who can help gets smaller, because a friend who is ready to lend you a few thousand dollars would hardly accept a mortgage on his house for the sake of helping you.

The financial aid that banks offer startups is usually relatively small, but the same can't be said of the *interest*. It is usually compound interest, so asking a bank for money is a rather expensive solution.

Also, you'll have to work hard to convince the bank's executives not only of the validity of your idea, but also of your ability to make it real. (Actually, this last point is vital regardless of the source you choose for funding.)

Alternative Solutions

There are many other funds you can access; let's have a look at some, starting from the most established ones and moving toward the new possibilities offered by the Internet and through virtual communities.

Local and Regional Economic Development

Many states and regions in the US have Economic Development Corporations (*http://bit.ly/1NyJBub*). These are usually either nonprofits or quasi-governmental agencies that can provide guidance, and in some cases early stage funding, for startups. Some states and regions may have specialized programs for funding technology startups. For example, Rhode Island's state-funded Slater Technology Fund (*http://bit.ly/1NyJFtT*) provides seed financing to promising startups and helps them make connections with other investors.

The New Angels

Angel investors are often retired company managers or entrepreneurs, with money available, wishing to invest in a young enterprise rather than in traditional financial instruments.

These angels usually come into play at the early stages of a company, driven by the challenge and the desire to be part of something new and rewarding from a personal point of view. Besides providing finances, angels give the project more credibility, and put their own contact network at the founders' disposal. They're also great sources of advice, leading the new entrepreneurs along the difficult path they are going to face.

Of course the economic aspect can't be overlooked, because angel investors usually expect a significant capital return. This rate is justified by the risk connected to the project, which for technological startups is often linked to non-tangible and hardly quantifiable aspects. Because the typical angel is not particularly interested in the ordinary management of a company, and is much more attracted by new challenges, he usually achieves the economic return by selling capital shares that he buys out against the investment.

Achieving funding through an angel investor is in some aspects easier than with banks and finance agencies, because the relationship established between founder and angel is very straightforward, based on mutual trust and shared values and visions. Therefore, the time needed to obtain funding is rather short, typically within months after the first meetings.

Venture Capital

Similar to an angel investor is the venture capitalist (VC), someone who invests high-risk capital, coming from funds that she manages, to finance startups expecting a very high rate of return, usually in technological and innovative fields. (Google is a perfect example of a venture capital success story.) Venture capitalists don't address the same market as the formal investors; they target companies with highly scalable businesses and returns of a bigger magnitude than the investments.

With a venture capitalist on board, there's a very high risk that the entrepreneur can lose the management of their own company. VCs usually invest millions of dollars, and investors tend to not leave the running of companies to inexperienced people. The VCs' main goal is not about having contributed to the launch of a new enterprise, as it may be for the angel investor. Rather, venture capitalists are all about recovering their investment quickly, with the sole aim of raising the sell-out price of their own company shares in the short term.

Crowdfunding

A recent phenomenon made possible by the Internet is crowdfunding, literally financing by the masses. This process allows an entrepreneur to present his own project on specialized websites, asking the community for financial support. The community might be made up of people who are interested in participating in the project development, potential clients, or just supporters.

This kind of funding has many advantages. First and most important, in order to launch a project, you must raise a certain amount of money by a certain deadline. If the money doesn't arrive by the deadline, you have a good indicator that your idea might not be as wonderful and revolutionary as you thought. There is nothing more wasteful then using resources (particularly time) to devote yourself to an activity that won't bear any fruit. So, if your ideas are rejected, try to understand how you can improve your offering and celebrate not having wasted your time.

Another advantage is that the founders keep full rights to the project, because it is not financed but *sponsored*: no one who donates to a crowdfunding campaign expects a financial return, or intends to take over the company. Crowdfunders typically expect to receive a product (or some keepsake related to the product, as shown in Figure 8-1).

Figure 8-1 *Components from KippKitts, Arduino, and Maker Shed await eager hackathon participants in Providence, Rhode Island*

There are several levels of involvement in a crowdfunding campaign, so it is possible to contribute donations of different amounts connected to more and more substantial rewards: a virtual thank you, a customized t-shirt, a mention on the packaging, access to the early versions, the chance of purchasing at a discounted price, one or more items of the finished product, and more.

The benefit for the sponsor is to have access to the product or service before others, sometimes even participating in its development. This kind of approach by all interested parties fosters the creation of communities around your product, creating a virtuous circle where the client turns into the first promoter of the project. Word of mouth is a very important tool here, because it is very effective and costs you nothing. The Internet takes this idea to the extreme, allowing some projects to grow until they become viral, often in a very short time.

In the United States, there are many crowdfunding platforms. The most popular one is Kickstarter, though we musn't forget others like Indiegogo, FundedByMe, or RocketHub.

Beyond Financing

What if you are not looking for financing? What tools could you use to make your activity flourish and turn it into a profitable business?

Bootstrapping

For those who don't want or don't like the risk of investors, there is an alternative: creating a business that can support itself starting from zero, or nearly so. Entrepreneur Chris Guillabeau claims it is possible to start a new business spending about a hundred dollars. Again, the Internet comes to the rescue with free or low-cost services: you can use Weebly or WordPress to create a website, MailChimp to manage your newsletters (including A/B testing experiments), and much more. Steve Blank (see "Success, This Time for Real" on page 49) maintains a list of these tools (*http://steveblank.com/tools-and-blogs-for-entrepreneurs*).

What's the Right Solution?

All the forms of financing and tools we have analyzed are not exclusive: they complement each other to provide a service that is tailored to the needs of different realities. We could start an adventure in bootstrapping for our first experiments, use crowdfunding to make a marketable product, ask a business angel for help with the creation of an actual company, and finally scale up with the contribution of venture capitalists. The different opportunities are typical of different stages of a company lifecycle and have to be regarded as such.

Always keep in mind that each enterprise has its own ideal size, depending on the type of product, service, and clients, and many other factors; growth is not necessarily the end-all and be-all.

Collaboration

9

In the past, the community of craftspeople was closed. Guilds, or collections of artisans, rigidly controlled who became a carpenter, a mason, a glassworker, a weaver, and many other trades. Master craftsmen were licensed by the guild, and only they were authorized to teach their trade, almost always through a long aprenticeship program. People who practiced a craft without being licensed represented a danger to the guild and had to be fought. Guilds pretty much died out in the late 1700s, although much of its nomenclature still remains ("journeyman plumber"). These days some craftspeople, frightened by the possibilities of globalization, demand the introduction of guild-like safeguarding systems instead of looking forward and modernizing their technologies and business models.

The Importance of the Net

Today, in place of guilds, there are specialized associations, which, except for a few cases, do not have a role of monopolistic protection but rather are more of a point of reference in the dialogue with institutions.

Makers are somewhat like artisans, when we consider the passion they put into their work, their desire to experiment and improve, their attention to details, and their habit of making do with and continuously inventing new uses for the available tools.

But the makers' approach to sharing information is radically different from that of guild craftsmen. Sharing for a maker isn't a bad thing. It is a powerful tool to help people reach their goals; *collective intelligence* is an unlimited and irreplaceable resource. We are not saying that a maker can't and shouldn't aim at creating brands and patents; makers simply recognize and accept alternative possibilities.

The Internet allows us to share our projects in a simple way. The fact that a click is enough to turn a digital model into a physical object is a very strong incentive not to build our own models from scratch, but instead to look for a preexisting model that is suitable for our needs, or easily modifiable. While looking for such a model, we frequently run into people with our same interests and same needs. Together with these people, we can create a community that turns a project around, such as MAKE's site (*http://makezine.com/projects*) or Instructables (*http://instructables.com*) (see Figure 9-1). The people we meet often have competencies that are very different from ours in terms of application, skills, experience, and culture. This variety is a crucial component of the great contribution a community can bring to a project.

The person who can best describe a problem is often the one with the best chance of solving it. Increasing the number of people involved in a project also increases the number of descriptions of the problem, likely bringing us closer to a solution. In addition, the mixture of competencies that different people bring to a problem makes mutual growth possible, gives everyone the chance to learn from the others, and improves the overall understanding of the current project.

For a maker, the creation path can feel more important than the finished product. For this reason, the possibility of growth during the project's execution is extremely significant, and the Internet can help us a great deal in this sense.

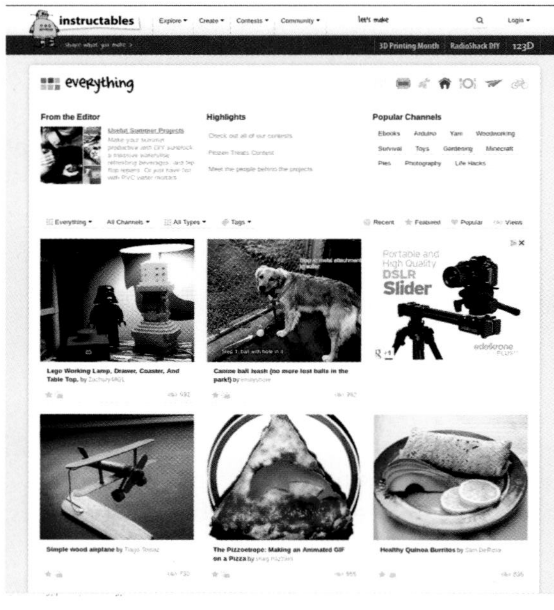

Figure 9-1 *On Instructables you can find instructions to build practically everything*

An Open Process

Making your work available to others can seem difficult. What if it is not good enough? What if people don't like your work? What if people like it and someone steals your idea? There are thousands of reasons that can stop you. We come from a long time tradition of patents, in which we are completely immersed: consider the continuous lawsuits between the giants of consumer electronics that sue (and pay) each other for infringing upon one anothers'.

Ideas are free; what actually counts is their realization.

Moreover, even with all instructions available, without a perfectly reproducible industrial process, the final result is not within everyone's grasp. Paco Torreblanca has no problem publishing bakery books with very detailed recipes, because even his best colleagues will never make his masterpieces.

But is it the same for the digital production process?

Indeed, a click is enough to allow anyone to manufacture an object completely identical to ours.

So why is it important to share?

Distributed Intelligence

Everyone likes to think he is unique. We *are* unique, but only up to a certain point. Even if—and this is a big "if"—you are the most brilliant person on your team, there will always be a great deal of people who are better than you at something. Even if you are one in a million, there are thousands like you in the world.

In certain types of maker communities, people aren't in it for the money, but for social capital in the form of reputation. Some maker communities are based on meritocracy, because it is the community itself that votes and supports the different ideas, without considering who proposes them (usually it is people we don't even know), exclusively on the basis of their value.

The production process can require many different competencies: electronics, mechanics, engineering, art, and so on. It is really difficult to be an expert in all these fields. But it is a good idea to step out of our "comfort zone" and at least become conversant with different competencies. We should try to concentrate on our strengths to best develop them, and live with the fact that not everyone can dance the *Swan Lake* ballet or sing *Nessun Dorma* before an enraptured audience. Instead, by taking part in a community, we can each give our best and the composition of different functionalities guarantees the harmony and quality of the final project.

Yes, right...but what if they steal our project?

A New Protection

Traditionally, the pattern has been to patent something as soon as you have invented it. A patent gives you exclusive rights to produce and sell that invention for a number of years; no one else can make money off of it without your permission. But lately, some people in the maker movement have concluded that this strategy—in a dynamic and distributed context like the one we have described—is not successful, for a number of reasons. First of all, when we start discussing our project in public, we still don't have anything completely definitive, so we are not able to patent it before exhibiting it to the public. In an iterated process of product development, we would need to patent the product several times—each time the design changes. Patenting an invention can be rather expensive, so this strategy quickly adds up, and the end result is a series of patents for unusable objects.

Moreover, if we're collaboratively designing something, who would own the patent? The volunteers who did the work? Someone else? This presents a huge ethical conundrum, which could damage that social capital we have built up so far. On Thingiverse.com (Figure 9-2) and YouMagine.com, for example, users are encouraged to upload their 3D designs, while other users are encouraged to remix those designs to create new ones.

The alternative strategy, then, is sharing one's own (if we can still call them this) creations. What's more, with some types of license the protection works in the opposite way: we can use what we have created on the condition that we continue sharing it so that other people can benefit from it, just as with open source software.

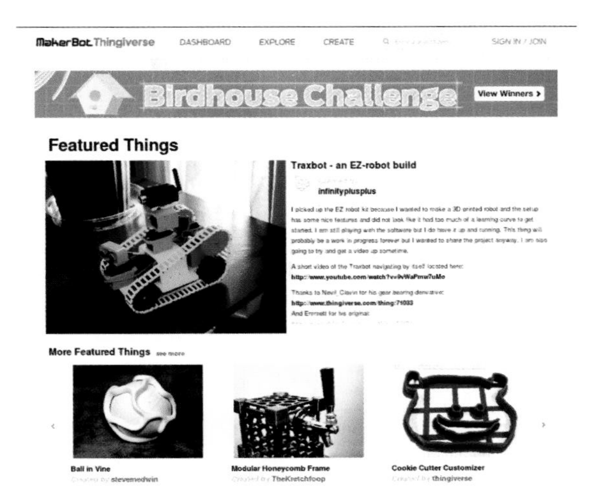

Figure 9-2 *Thingiverse: a universe of objects*

Creative Commons Licenses

The Creative Commons (*http://creativecommons.org*) licenses allow us to specify the ways in which one of our creations can be used.

There are four conditions of use that we can combine according to our needs:

- Attribution required (*BY*)
- Using the work only for noncommercial purposes (*NC*)
- Prohibiting the creation of derivative works (*ND*)
- Distributing derivative works only under a license identical to the license of the original work (*SA*)

If, for example, we wanted to permit the distribution of our work for commercial aims, too, as long as no modification is made we could use CC BY-ND.

What can we do, then, to have the authorship of the project acknowledged? Once more, it is the community we have built in a climate of mutual esteem and trust that helps us, acknowledging us as the legitimate authors. The participants not only help us, but they also represent a direct return in terms of visibility, fairness, and reputation. In an indirect way this increases our network of acquaintances, helping us find better opportunities and offering us a new springboard for our personal and direct growth, and our economic growth too: who wouldn't like to work with one of the leading personalities of a successful project at an international level? Even if their main driver is passion, this is another reason why makers continuously look for interesting and promising projects that can attract people's attention and gain approval.

It is true that with this approach we don't get the level of legal protection afforded by patents. In its place, there is a project created with distributed intelligence, for which all marketing is made by the community itself, composed at the same time by technology evangelists, designers, testers, people dealing with documentation—in other words, people who carry on the project as if it were theirs. Well, in a way, it *is* a bit theirs too, so it is more of a mission than a job, and they carry out this mission with passion and performance rarely seen in an employee. In addition, as they are not only our collaborators but also our first customers, they pay us.

Bits, Bytes, and Atoms

By operating in this context, we share anything digital, thus permitting the reproduction of our artifacts on the other side of the world too. The only thing we keep is the brand, which allows us to unequivocally identify an object coming from our productive pipeline, just as with open hardware products such as Arduino.

Sure, everyone can build an Arduino-compatible board, but few people will actually do it. Most people will prefer buying it directly from the manufacturer, who offers the bits and bytes for free (all necessary information to create the product) but who sells the atoms (finished product). Considering what we have said so far, the preferential manufacturers will be the ones who have created and supported the project rather than an unknown clone of it. The clone may have its own market too, but if you consider the figures of the various manufacturers, you'll realize the validity of this heuristic process, also because the clone won't be nearly as successful in terms of distribution and commercialization. The important point is not to fall prey to forgery—the only true offense to this business model, because it doesn't protect the original manufacturer or the consumer. For example, although you can make an Arduino clone, you can't call it Arduino without licensing the name because it is a trademark. However, should a clone introduce some improvements, the whole system would benefit from it, because other realities may start again from the modified project for a further innovation.

But what if they really steal your market? Maybe you need to take a look around and start with something new. We may then end up creating a product or a service with a higher surplus value. Maybe you'll be inspired to create the next generation of your product, just as Arduino has done, from the original Arduino to the Uno, Leonardo, Due, Tre, Mega, Yún, and many more.

In this case too, you benefit, as does the whole system.

PART III

From Bits to Atoms

God took the mud, spat on it, and Adam was born. And Adam, wiping his face, said: "Good start..."

—Giobbe Covatta,
Parola di Giobbe

- Chapter 10, Managing Project Files
- Chapter 11, This Is Not a Pipe
- Chapter 12, 3D Printing
- Chapter 13, Milling
- Chapter 14, Laser Cutting

Managing Project Files

10

A current-day maker, unlike the artisans of the past centuries, doesn't spend time locked up in his workshop far from prying eyes. The maker paradigm (and business model) is all about sharing: sharing ideas, sharing designs, sharing code. Since nowadays nearly all that information is digital, we are really talking about the management of project files.

How can you keep control over a project when so many people can take part in its development? You need to establish some kind of process that allows for the evolution of the product, preventing the project from devolving into chaos. Fortunately, there are tools that can help.

Distributed Design

Probably the first solution that comes to mind is to maintain a revision of your project files designated as "the active version," while still maintaining all previous versions as backup copies. The project manager, or someone specially assigned to the task, collates all the inputs, archives all the old versions, and declares the latest iteration to be the active version.

It sounds pretty simple, but this method is awfully prone to mistakes. Contributors sometimes make changes and forget to document them, or the project manager might miss a critical update. Even if all the records are kept in perfect order, and all the backups are maintained, it can still be difficult to track a change from its origin. Who gave the order to change all the ball bearings to cubes? You'll find the answer somewhere in that nest of archives.

If this sounds familiar, don't worry: you're not alone. Other people have had the same problems, and they've come up with quite a number of solutions.

In the software world, managing all the contributions from the community is an old problem that has been solved with a number of tools that allow a manager to keep control of all of the different versions of a project. This is generically called a *version control system* (VCS).

Even though VCSes were created for software, we can use them for any kind of file—for example, to manage the chapters of a book.

These systems allow for clear tracking of all changes made to models, documents, source code, or other project data so that we can follow its development throughout time. Hence with a VCS it is possible, and easy, to restore a specific version of a particular file, or even of the whole project. You can use a VCS to recover data that have been deleted by mistake. You can track how a specific problem in the project was identified, diagnosed, and solved. You can also try out alternative solutions, without running the risk of permanently damaging the project.

The early version control tools were centralized (Figure 10-1); that is, there was one single repository that kept track of all project files, whereas each client (the computers of people participating in the project) kept only a copy of the most recent version. They worked fine, and products like CVS and Subversion, which were two of the most popular VCSes in the industry, still work that way.

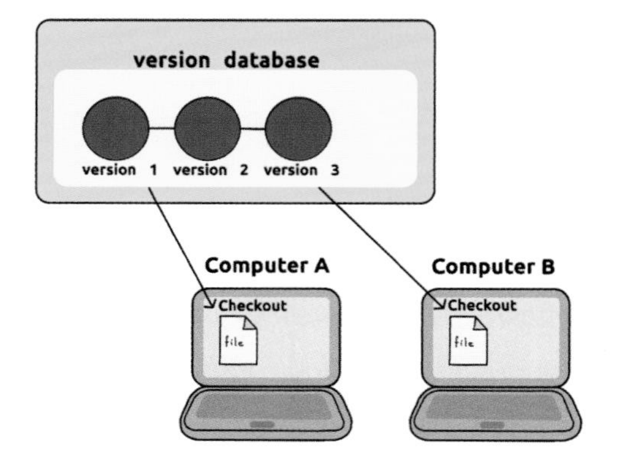

Figure 10-1 *A centralized version control system*

This system proved to be a little restrictive, so in recent years *distributed version control* tools (Figure 10-2) like Mercurial and Git have gained popularity. The latter, together with the online interface GitHub, seems to be the most popular and beloved by the community, as of this writing.

In a distributed control system, even if all our data were destroyed and we didn't have a backup copy (which is surely not the case...right?), we could recover them from a colleague's data, without going crazy tracking the history of every file. The software will do that for us.

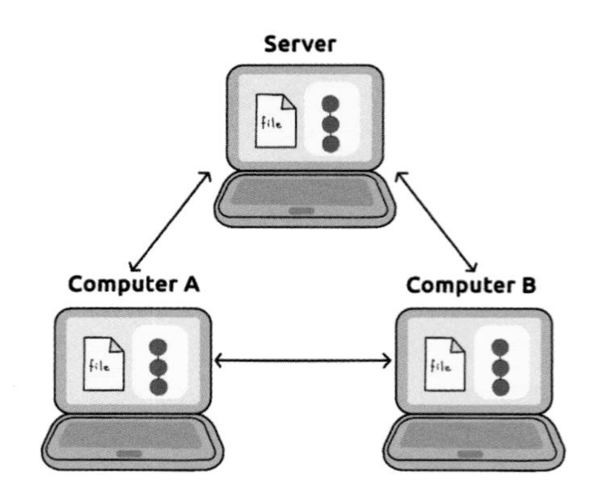

Figure 10-2 *A distributed version control system*

Git and GitHub

Git was created to meet a project's distributed design needs, though any individual can benefit from the great power of this tool as well.

We won't go into depth on how the versioning software manages its files and data; all we need to know is that each repository knows the exact internal status of the document at any moment, as if it took a picture of each single file, as shown in Figure 10-3.

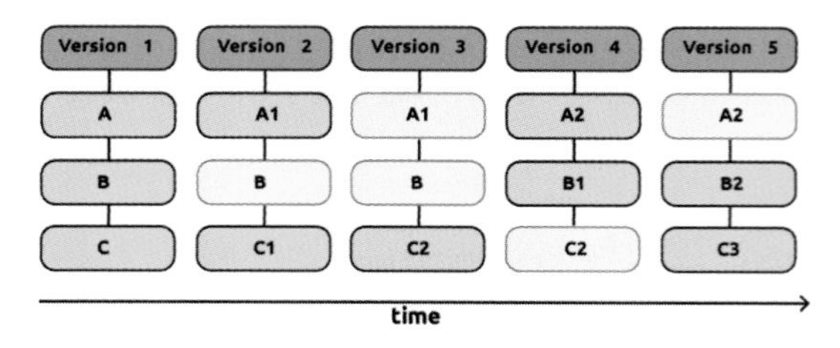

Figure 10-3 *Git keeps a series of snapshots of the project throughout time*

Creating a New Project

GitHub puts at your disposal a totally free system for all open source projects. Actually, GitHub is a *freemium* service: if you want to use GitHub for private projects that aren't open to the world, you need to pay; however, the software is free, so you could always set up your own private Git server for no cost. You only pay a premium for hosting private workspaces on Git's own servers. In the previous chapters we tried to convince you that sharing is a winning choice, so we'll show you how to use this service to host the main version of your project's repositories, open to the public.

Before starting to use GitHub you need to create an account. Open your browser, go to *https://github.com*, enter your email address, choose a username and password, and click on the registration button, as shown in Figure 10-4.

After registration, access your main page, where you can create a new repository by clicking the "plus sign" icon, at the top of the page next to your username, and choosing New Repository, as shown in Figure 10-5.

On the page that comes up, enter a name for the repository. A good rule of thumb is to choose a short, easy-to-remember name. If you are short on ideas, you could take the project name and turn it into camel case by grouping the original words together, with capital initial letters, into one name (e.g., MyProjectName).

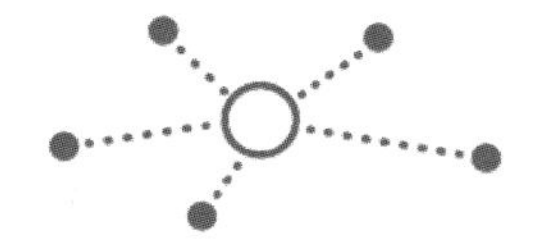

Figure 10-4 *The registration page on GitHub*

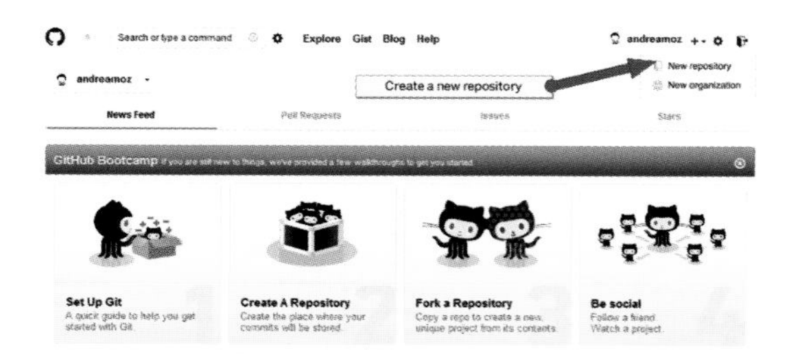

Figure 10-5 *The GitHub main page*

If you'd like, you can also enter a short description. If you check the box labeled "Initialize this repository with a README," you will be able to *clone* the repository later—that is, create a full copy on your computer.

Click the "Create repository" button (Figure 10-6), and you have created your first repository!

Now you are ready to start working (Figure 10-7).

The Three Areas of the System

The work that you perform in Git exists in one of three states, shown in Figure 10-8. It can be *modified*, *staged*, or *committed*. Each stage is different from the others, and understanding them is important to smoothly use Git.

Modified means that you have performed a change on some file in your project, but you have not yet committed those changes to your database. A modified file exists in a kind of limbo; it is no longer part of your project's history, but it is not yet part of your project's future.

A staged file is a modified file that has been marked to go into your project database the next time you commit your work. This is your way of telling the system that the file will be part of your project's future, but not yet.

A committed file is a staged file that has finally been safely stored in the project's database. From now on (until you or someone else changes this file) this is the "official" version of the file.

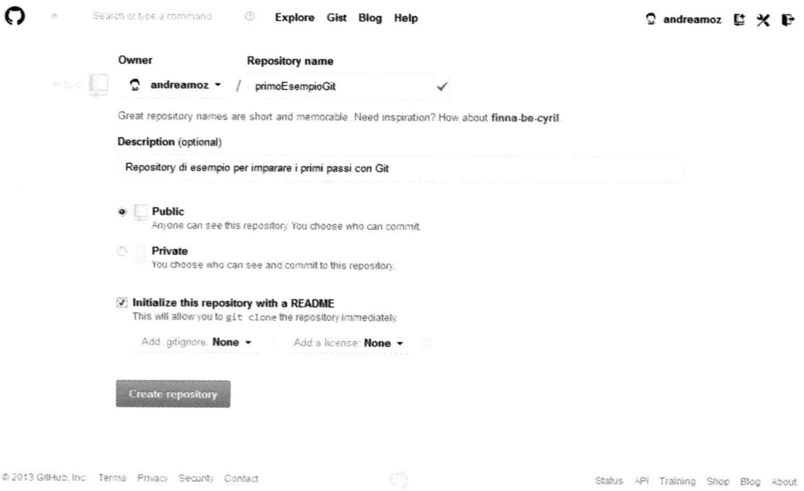

Figure 10-6 *Creating a new project*

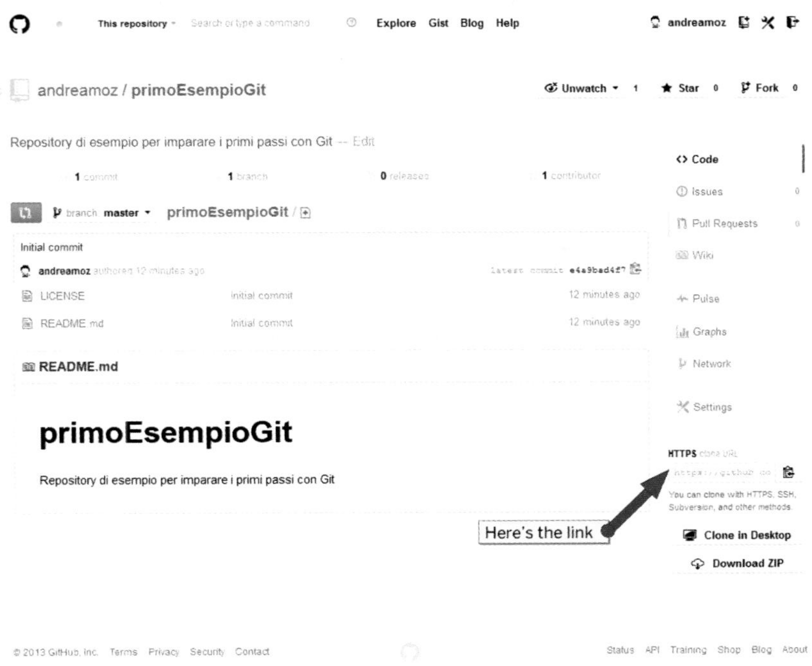

Figure 10-7 *The new project, ready for use!*

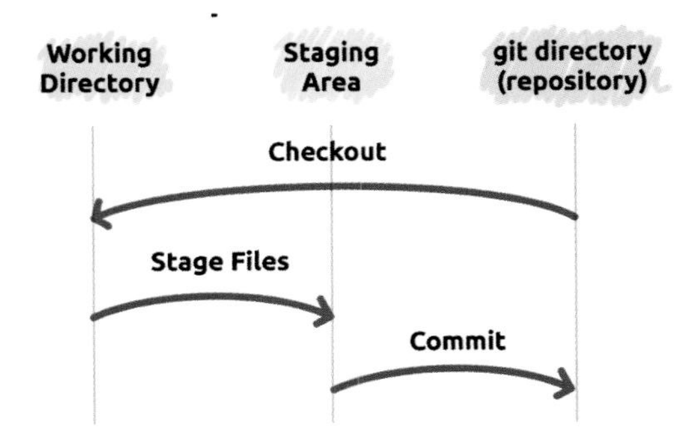

Figure 10-8 *The three states in local operations*

The basic workflow is rather simple:

1. Download the documents from the central Git repository into the working directory on your computer.
2. Make all the necessary changes.
3. Put all the files you want to store in the repository into the staging area.
4. Perform a *commit*, the operation that finally saves the files to the history of the project.
5. Perform a *push*, which pushes your local project history and state to the central repository.

Installing Git Locally

Please refer to the Git website (*https://help.github.com/articles/set-up-git*) and the official guide (*http://git-scm.com*) for the most current instructions.

Git can be used with a text interface—called a *command-line interface*, or CLI—or with a graphical client. The CLI gives you more direct access to Git commands, and is the first choice of many people who work on software professionally.

There are many graphic Git clients available for all the most popular operating systems. Let's see how to use Atlassian SourceTree on a Windows system. It's also available for Mac.

We first need to download the software (*http://www.sourcetreeapp.com*) and install it, leaving the default settings. After the installation, launch the application.

The Workflow

The first step is to clone a copy of the project on your computer: click the icon labeled Clone/New or choose File→Clone→New on the menu. Paste the public address of the repository, as shown earlier in Figure 10-7, in the Source Path/URL field. In the Destination Path field, type the complete path of the location where you want to save the project files on your computer.

Click the Clone button (Figure 10-9), and after a short time you'll have a full copy of the central repository on your computer. In the central bottom area of the screen you can see the project files, at the moment restricted to an (empty) license and a text file with the introduction message you opted for upon creating the repository in GitHub (Figure 10-10).

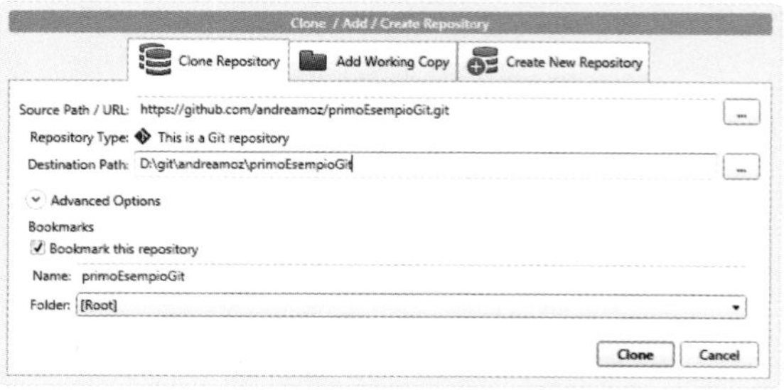

Figure 10-9 *Cloning your first repository*

Now, you can edit the *README.md* text file with any text editor. Add the line "This has been changed for sourcetree," and then save the file and close the editor. When you go back to SourceTree, you'll see that SourceTree noticed that you made some changes: a new notice has appeared toward the center of the screen, reading "Uncommitted changes." Click on those words, and at the bottom right of the screen, you can see exactly what has changed (Figure 10-11).

The edited lines are colored green to show they have been added. If we had deleted lines, they would be still be visible, but highlighted red. This feature is not available for binary files, such as images or videos.

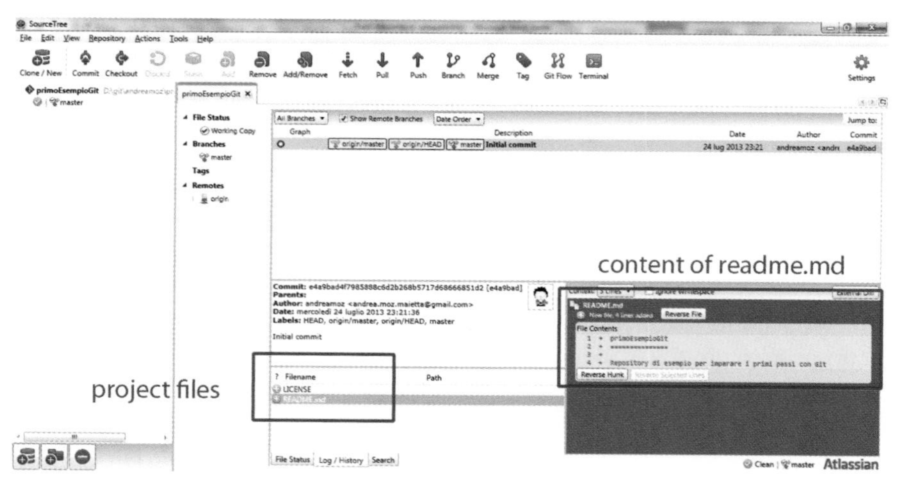

Figure 10-10 *The repository on your computer*

At the moment, the change you made to the *README.md* file exists only in your working directory. To move the file to the staging area, you need to click the Add button or choose Actions→Add from the menu.

This action moves the *README.md* file from the Working Copy Changes section to the Staged Changes section. If you had edited more files, they would all have moved from Working into Staged when you clicked Add.

Now you are ready to transfer the edited *README.md* file from the staging area to your local repository, as shown in Figure 10-12. You have to click the Commit icon on the toolbar, and type a message that explains the reason for the commit so the other people working on the project know what you did (or so you will know when you look at it later). Describing the individual changes made (such as "inserted 'word' at line 5") makes little sense; Git makes it easy for the other people in your project to see what you've done. The commit message is where you tell them *why* you did it.

A good commit message shows a high-level view along with the reasons for the change, like "Split the subscription plans into 6 because of the phone call with Alice." While the first line should be 70 characters or less, you can add newlines and more paragraphs; having some keywords in a commit message makes finding this specific commit later on much easier. When you've finished explaining what you did, click the small Commit button at the bottom-right side of the screen (Figure 10-13).

When you click the Commit button, SourceTree tells you that the edited file has been put in the local repository. On the toolbar, the Push icon now shows a 1. This means that the changed file is not available in the remote repository yet (i.e., on GitHub), but is instead ready to be pushed there (Figure 10-14).

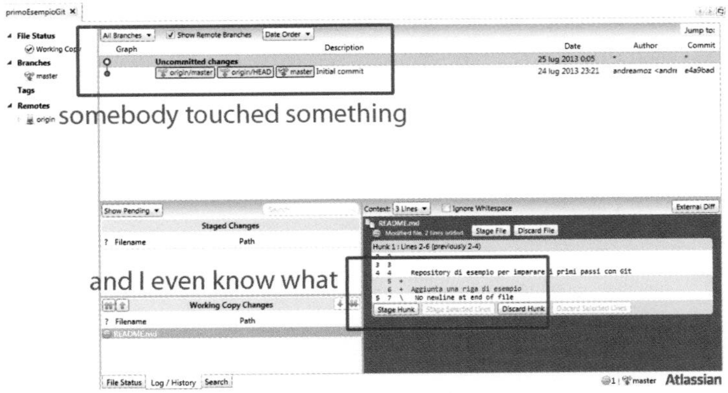

somebody touched something

and I even know what

Figure 10-11 *Nothing gets past Git's watchful eye*

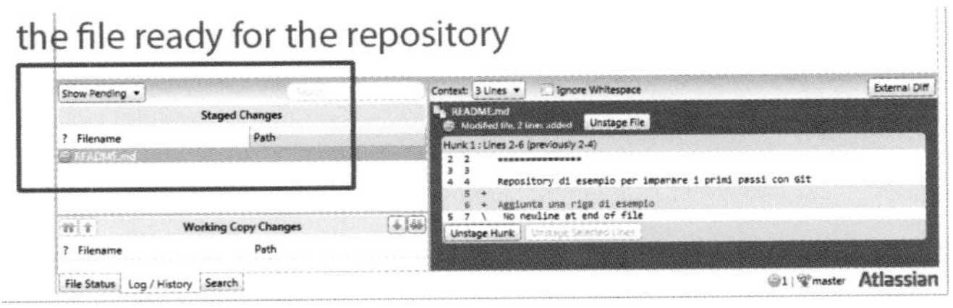

the file ready for the repository

Figure 10-12 *Changes ready to be transferred to the local repository*

To copy the changed file to the remote repository, click the Push button. You are shown the address of the repository you cloned, which by default is named *origin* (Figure 10-15).

If you now click OK to end the operation, you get a message asking you to input your GitHub username and password. You don't want anybody to be able to edit your data without permission, right? Type both in and click Login, and your changes will be on their way to the server!

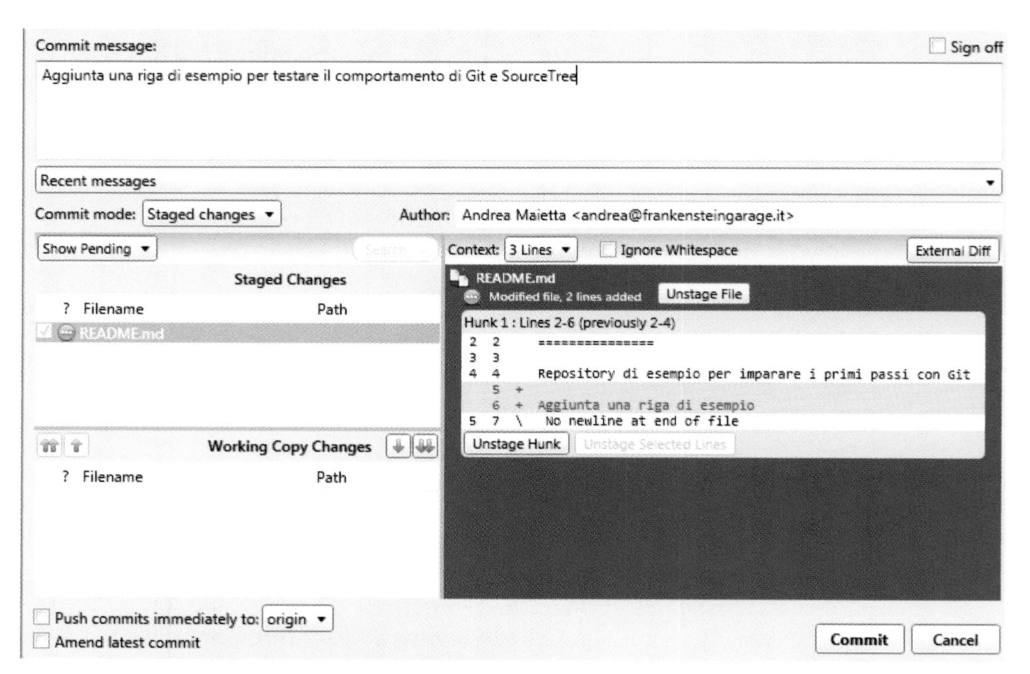

Figure 10-13 *Your first commit*

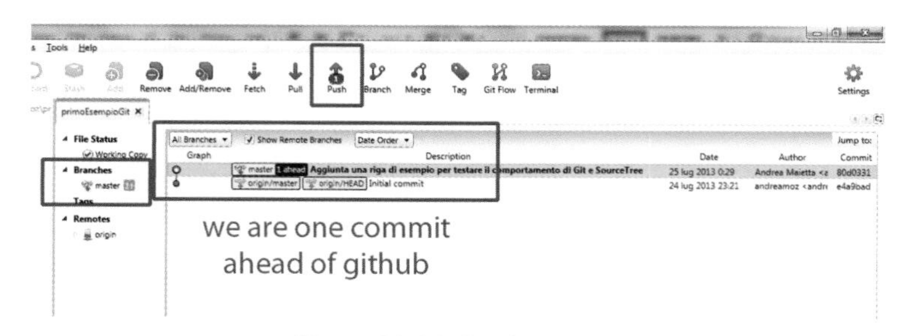

Figure 10-14 *Ready to push*

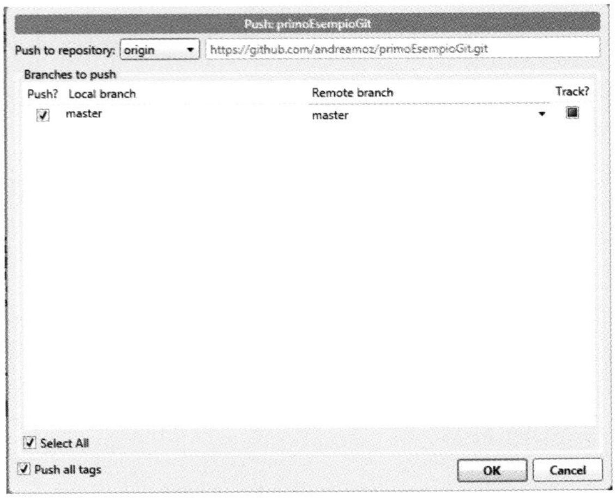

Figure 10-15 *Sending the changes to the remote repository*

You are done, finally! Now SourceTree is happy because you have synchronized your changes to the remote repository (Figure 10-16).

To check if the change has been transmitted, you can open the file in GitHub (Figure 10-17).

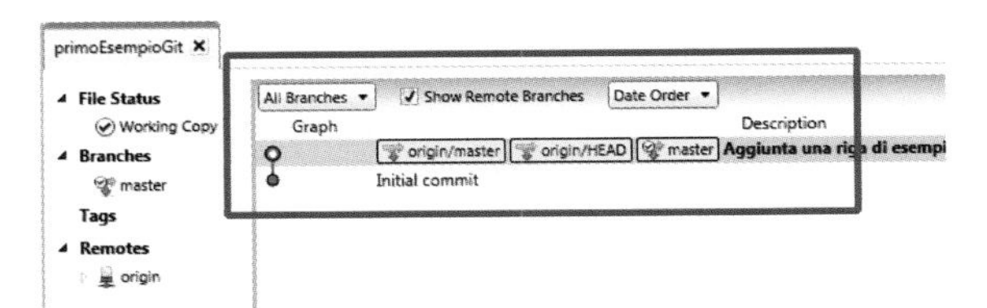

Figure 10-16 *Now we are in sync with GitHub*

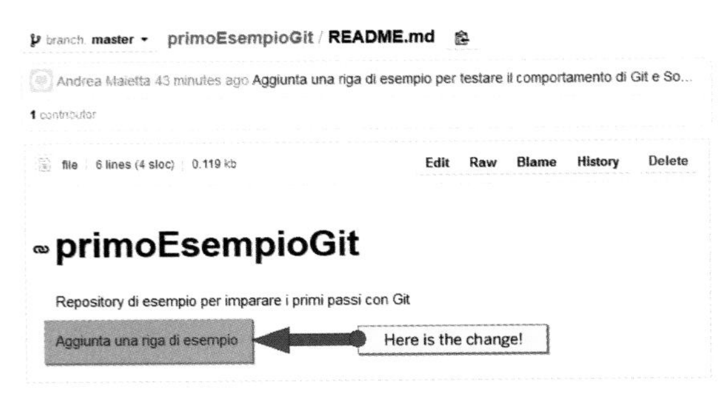

Figure 10-17 *Ensuring the change is visible on GitHub*

Let's try this action from the other direction. You're going to make a change to the *README.md* file in GitHub, and download the change onto your computer. The reason for this is simple: if you've got a number of people working on a project, some of them will be editing a file via one method (say, directly into GitHub), while others will be working offline in SourceTree and uploading their changes later. Before you start working on that same file, you need to synchronize your computer with all those remote edits.

Staying in GitHub, click the Edit button on *README.md*, then make another edit to the file. It doesn't really matter what you do, you just have to change it in some way. Click the Commit Changes button.

Then go back to SourceTree and click the Pull icon on the toolbar. When the Pull window opens, click OK (Figure 10-18).

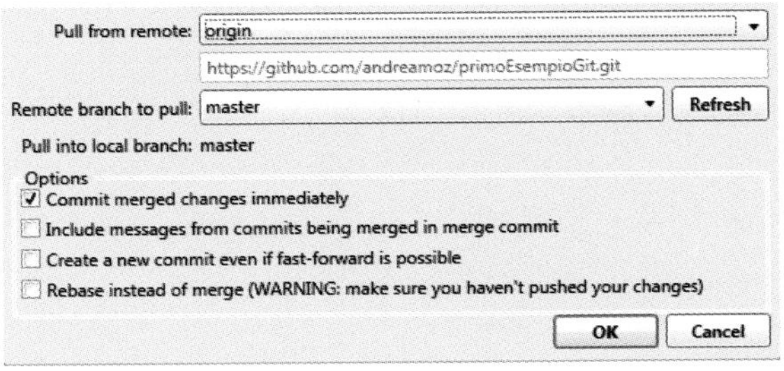

Figure 10-18 *Pulling from GitHub*

After you do this, your local repository will be synchronized with GitHub again, and you will be able to see locally the changes you have made in the remote repository. If you forget to synchronize your database before starting work, nothing serious would happen. However, the chance of a conflict—that is, different people editing the same thing in the same file—would increase. In this case, GitHub would show both versions, leaving you to choose the correct one. So...don't forget to synchronize often!

Not Only Trees Have Branches

Throughout the lifecycle of a project, there may be several instances where you must choose between alternative development paths. Say, for instance, you're building a scooter that'll come in two models: one that runs on fossil fuels, and one that runs on electricity. The bulk of the scooter's design—the frame, the seats, the brakes, the tires, and so on—will be exactly the same for both versions. The only differences will be the powertrain. Git can help you control the different versions with its *branch* feature.

In Git, a branch is used to develop an isolated alternate version of a project. The default branch—the body of the scooter, with a gasoline engine—is named *master*, as shown in Figure 10-19.

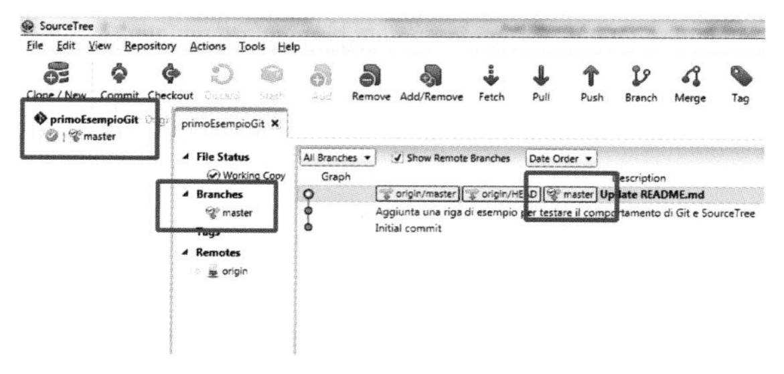

Figure 10-19 *Which branch are we in?*

You must first create various files representing the different parts of the scooter. You can see how SourceTree notices that you've added new files (Figure 10-20).

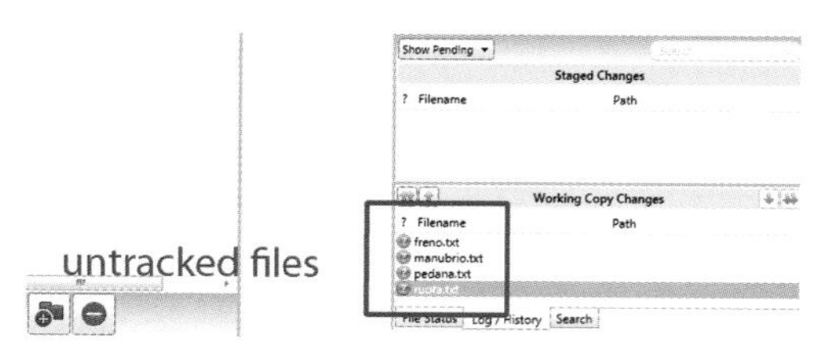

Figure 10-20 *New files added to Git*

As you did before, you must first Add and then Commit. Next, Push to synchronize GitHub and let your collaborators see the most recent version of the project. Now you have your scooter in its "standard" version.

Now you can work on its powertrain. Instead of duplicating the project folder, create a branch: click the Branch button (Figure 10-21), type in a name, and click Create Branch.

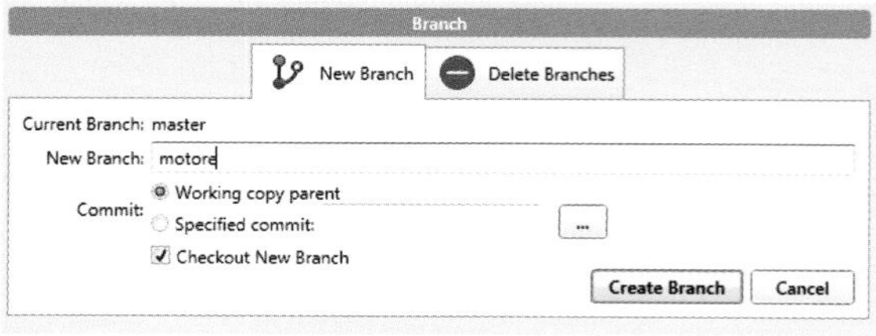

Figure 10-21 *Create the branch*

We can see that, in the Branches section, there are now a master and a powertrain folder, which is the branch we just created. The latter is flagged, because it is where we are now (Figure 10-22).

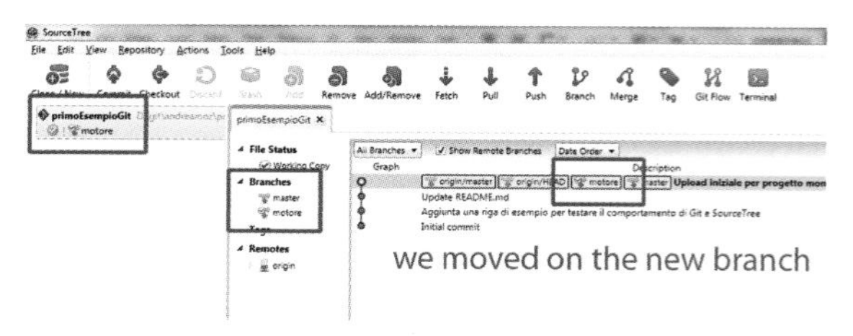

Figure 10-22 *You are in the new branch*

Now you can edit the project. Create a new file for the powertrain data and modify the files to create an electric version of the scooter. You can see that the two branches, which were identical at the time of creation, will proceed on separate paths until you decide to merge them (Figure 10-23).

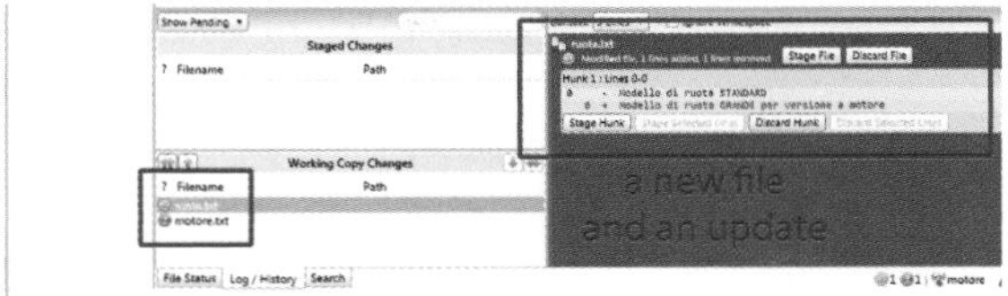

Figure 10-23 *The changes for the powertrain scooter*

As usual, you need to Add and Commit. In order to see all the files in the working directory, click the File Status tab, at the bottom center; you can see that the powertrain branch comprises the new electric engine model (Figure 10-24).

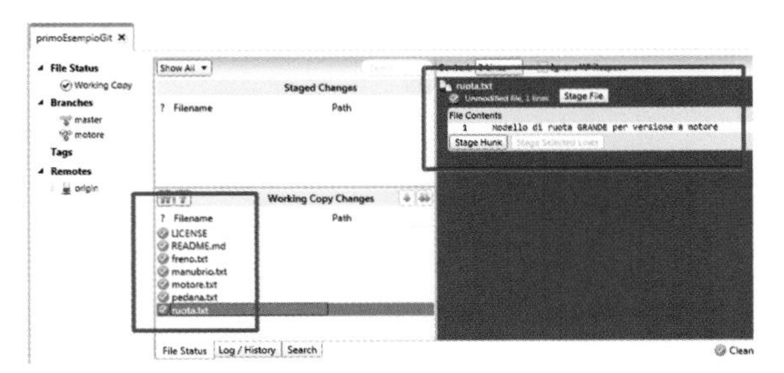

Figure 10-24 *The powertrain branch contains the new model*

To go back to the main branch, double-click Master under Branches; we can see that the design is for a gasoline-powered scooter, meaning that the two branches are actually distinct (Figure 10-25).

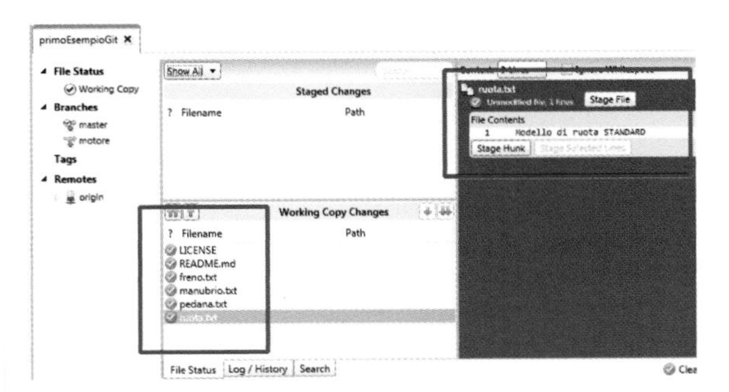

Figure 10-25 *Back in the main branch, there's no trace of the powertrain version*

To join the two branches, click on the Merge button, then select the branch that you want to merge with the current branch, and click OK. Figure 10-26 shows the merge.

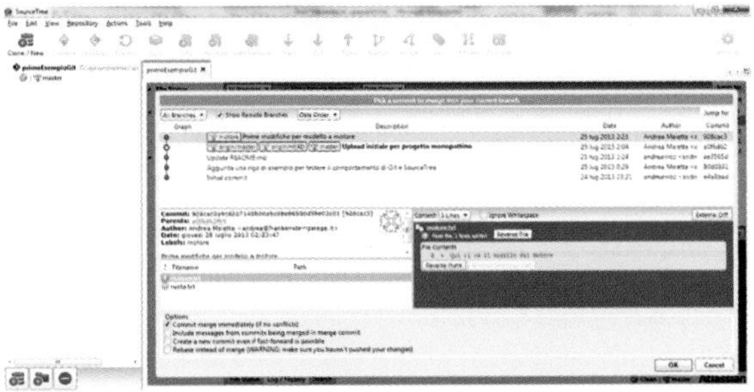

Figure 10-26 *Bringing two branches together with a merge*

With this process, everything is much simpler than archiving versions manually, because Git manages all the data behind the scenes.

We have explored only a tiny part of Git's potential, though it can already meet your most common needs. Most of all, you know enough to download a project and use it as a starting point for your own work.

What are you waiting for, then?

Let's start making something now!

11 This Is Not a Pipe

The bits-to-atoms revolution is bringing machines able to create real objects into our homes. These machines take minutes or hours to print a three-dimensional model, turning a file on a computer into something physical we can touch with our hands. The techniques used to create objects are many: we can deposit material, dig chunks out of blocks, and create molds to be filled or forms on which materials are laid. We can create these objects by ourselves or with a community, and we can share them, download them from the Internet, modify them, and create them in our homes. It's all about turning bits into atoms.

Manufacturing Processes

The technologies we are going to show you belong to two big families that existed long before the birth of the maker movement; they are just more accessible now.

When we were children and started playing with blocks, we learned that we could create objects by stacking or tacking together various pieces, layer after layer. In fabrication, this type of process is called *additive manufacturing*, because we add one material to another until we get the complete object. Some kinds of 3D printers work this way. An extruder, sort of like a high-tech glue gun, melts a plastic material and deposits it, layer by layer, according to the computer model, until an object is created.

A sculptor uses a chisel and other tools to carve marble or wood, removing excess material and leaving behind a beautiful statue. In the maker world, computer numerical control (CNC) machines use a rotating tip similar to a dentist's drill to grind away at blocks of wood, foam, resin, and even metals, leaving behind a sculpted object. This type of process is called *subtractive manufacturing*, because it subtracts material from a solid block of material.

Die-cutting machines also remove preexisting material, but they work only in two dimensions. They cut sheets of material using sharp blades, lasers, or high-pressure fluid. Also, with these machines makers can create 3D objects by combining different parts, more or less like we do with wooden kits to build a dinosaur's skeleton.

We can also combine the two techniques: for example, with milling machines and printers it is easy to create molds into which melted material is then poured slowly, or to create outlines to cover with papier mâché or fiberglass.

Starting from Bits

So you've got a design for a product. You've doodled it on napkins, and you've sketched it beautifully into a notebook. Now you need to turn that design into bits by converting your idea to a digital model.

To do this, you'll use something called Computer Aided Design (CAD) software. Such programs let you sketch your design directly into a computer, in three dimensions. While many CAD programs store their files in a proprietary format, there are some standards, such as Standard Tesselation Language (STL). In this format the objects are approximated with a series of triangles called a *mesh*; the more triangles there are, the better the model definition.

Software

One commercial software package appreciated by designers is *Rhinoceros*, also known as Rhino (Figure 11-1). It is easy to learn, can manage dozens of file formats, and can also be used for parametric design—the practice of letting algorithms influence design. With the Grasshopper plug-in, it can also create generative art.

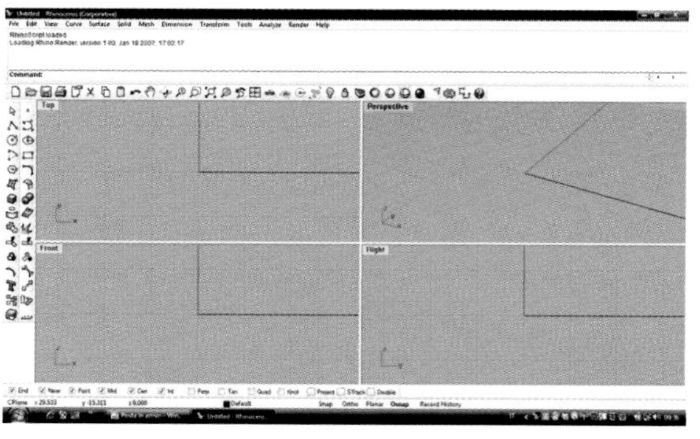

Figure 11-1 *A particularly light rhino*

Blender, shown in Figure 11-2, is an open source design software that is also able to carry out many operations beyond 3D modeling, like fluid simulation. Even though it was not specifically intended to create technical drawings, it is very good for creating files in STL format.

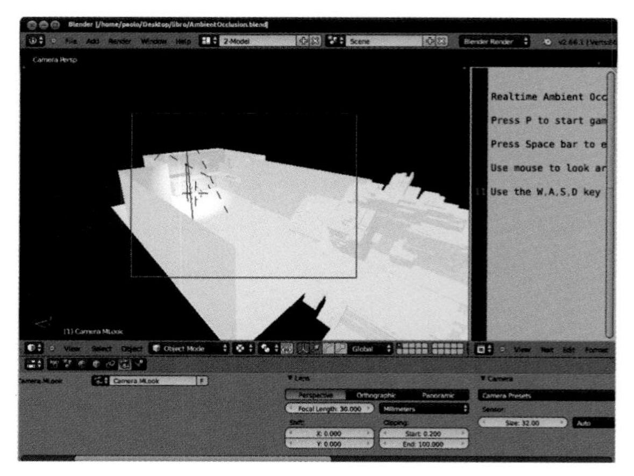

Figure 11-2 *Blender is quite complex but it is also very powerful*

The *123D Design* software (Figure 11-3), available online, belongs to the 123D suite by Autodesk. It is used to design printable objects in a simple way.

Figure 11-3 *Autodesk 123D*

Another package that is free of charge and quite simple to use is *SketchUp Make*. (Also available, for a fee, is the SketchUp Pro version.) It doesn't create files in STL format directly, so you'll need some plug-ins to convert the files, which makes the whole operation a bit more complex. It's shown in Figure 11-4.

Figure 11-4 *Trimble SketchUp Make*

A slightly different software package is *3DTin* (Figure 11-5), which runs directly in browsers such as Chrome or Firefox. It is free to use, but the models created must be shared with a Creative Commons license. It can directly export files in STL format.

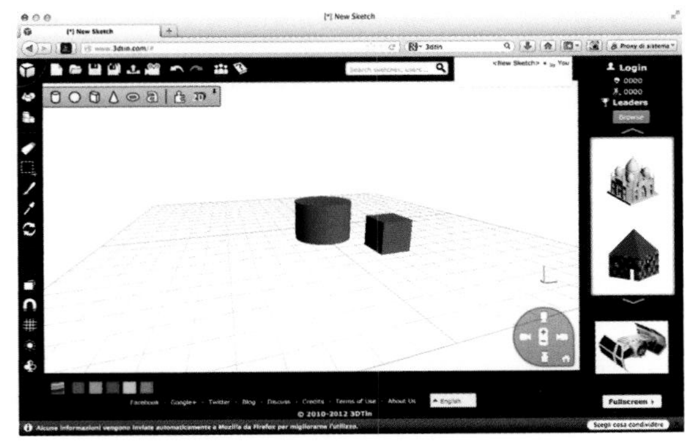

Figure 11-5 *3DTin is an online CAD application*

TinkerCAD (Figure 11-6) is an online editor too. The basic version is free of charge for personal use, but not for commercial purposes. It has a clean and simple interface and works more or less like 3DTin.

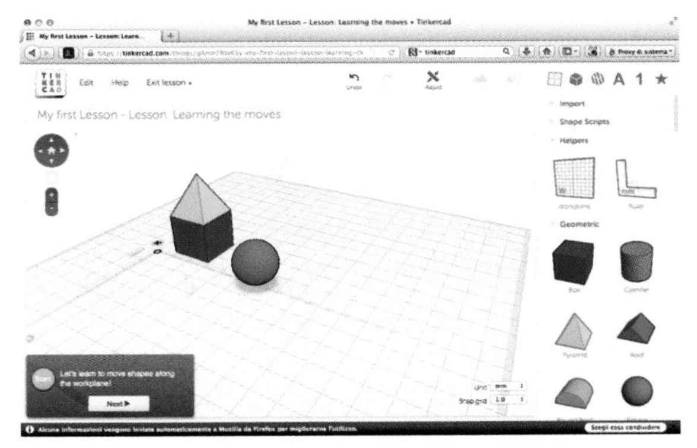

Figure 11-6 *TinkerCAD*

OpenSCAD

In all the software we just listed, most of the input is provided via the mouse. This can be a bit confusing at first, because you need some time to get the hang of how to operate in three dimensions on a flat two-dimensional mousepad. With Blender, which is a wonderful and extremely powerful software, by just moving an object you run the risk of getting lost in the buttons, options, and menus of the extremely complicated interface. For our purposes in this book, we have decided to start with *OpenSCAD*, a free software package that may look a bit spartan (Figure 11-7) at first, but is powerful and easy to use.

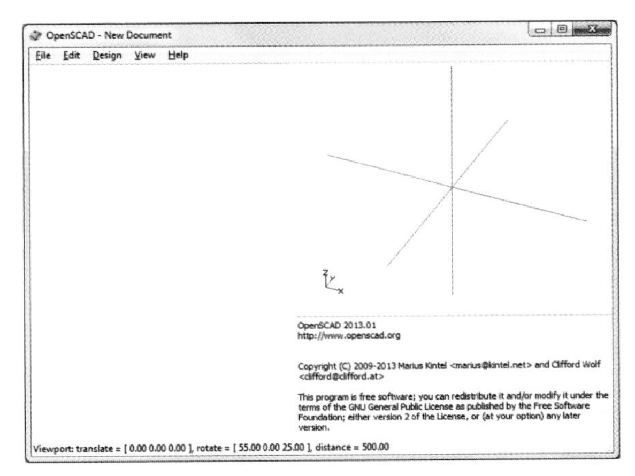

Figure 11-7 *OpenSCAD opening screen*

There are no complex combinations of buttons, mouse movements, or interfaces to learn with OpenSCAD, but it has a quite different approach: it uses a very simple language to write *scripts*—that is, sequences of instructions. OpenSCAD compiles the scripts and shows you a 3D image of the resulting object.

The interface has only three panels:

- An editor in which the instructions for the design are inserted
- An area to visualize the object in 3D
- A console for system messages, errors, and advice

With OpenSCAD the mouse works only on the 3D viewer window and lets you do only three things:

- Move the mouse, keeping the left button pressed to rotate the visualized object
- Move the mouse, keeping the right button pressed to shift the scene
- Use the mouse's central wheel to zoom in or out

Simple, isn't it?

Before starting, in the rightmost window you can see three perpendicular straight lines—the x-, y-, and z-axes—meeting at a point called the *origin*. You can define the position in each point of the space keeping these axes as a reference: the *x* coordinate measures the distance of the point from the plane formed by the straight *y* and *z* lines, called *yz*; the *y* coordinate measures the distance of the point from the *xz* plane; and the *z* coordinate measures the distance of the point from the *xy* plane. The origin coordinates are (0, 0, 0).

Hello, World!

Now, create a cube by writing the following instruction in the left box:

```
cube([1]);
```

Press F6 to make the software carry out your first script, and there you have it: a cube that's 1 mm long on each side! The image is a bit small, so use the mouse wheel to zoom in if needed. (See Figure 11-8.)

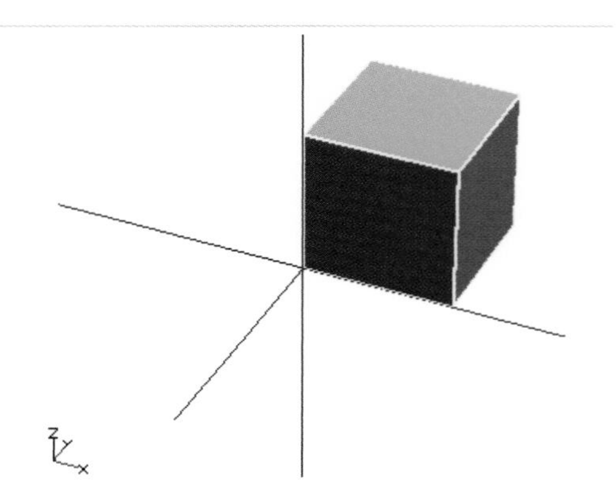

Figure 11-8 *A simple cube*

Now you could also export your model in STL format to then create it with a 3D printer. To do this, choose the Design menu and click "Export as STL." Name the file and save it.

With the View menu we can choose the point of view for our object, or choose between perspective and orthographic view.

Let's do some more experiments with the cube control. If you use three different values, you'll get a parallelepiped:

```
cube([2,3,4]);
```

The three numbers in the square brackets represent the length of the three sides of the parallelepiped.

OpenSCAD draws figures by always starting with a vertex on the origin of the axes. You can add a parameter and center the figure exactly in the origin:

```
cube(size = 5, center = true);
```

Beyond Cubes

Let's try some other figures. To make a sphere, use the following instruction:

```
sphere(10);
```

The number in parentheses is the sphere radius. As opposed to the cube, whose origin is one of the vertexes, the sphere is centered at the origin. Figure 11-9 shows the rendered sphere.

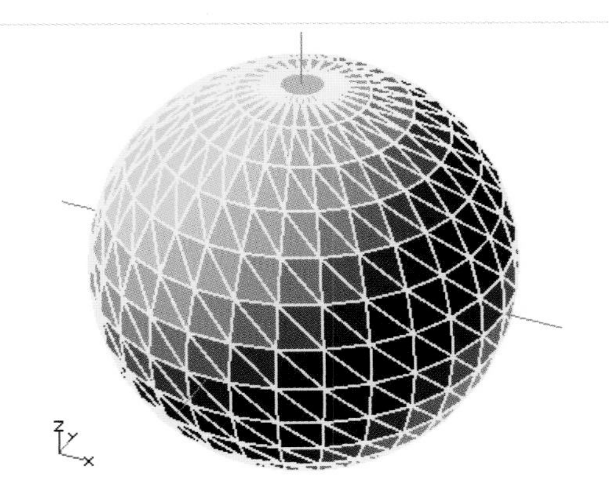

Figure 11-9 *A multifaceted sphere*

This sphere looks a bit too...squarish. If you were to print this object, you probably wouldn't be very happy with the result. It's a good thing you can modify the resolution by adding the `$fn` parameter:

```
sphere(10, $fn=100);
```

Now the resolution is definitely better (Figure 11-10), although we haven't reached perfection yet.

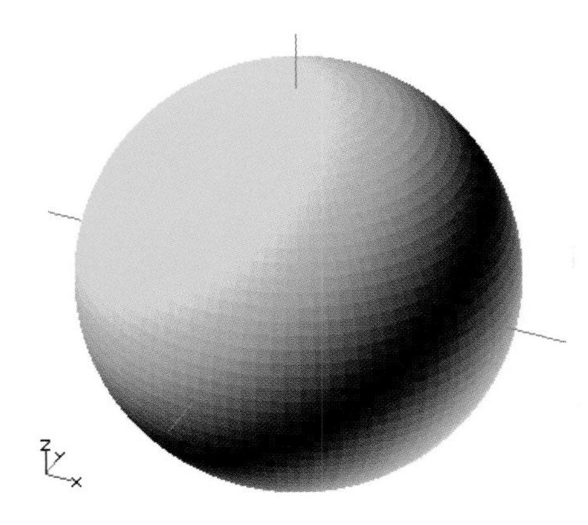

Figure 11-10 *The sphere with a higher resolution*

Continue experimenting; set the resolution at 3 or at 1,000 or any other number. In all of the following examples, and whenever you use OpenSCAD for your designs, you can choose the most suitable solution for your goals. After the sphere, we'll try a cylinder (Figure 11-11):

```
cylinder(h = 10, r1 = 10, r2 = 10, center =
false);
```

The first parameter is the height, the second parameter is the radius of the top of the cylinder, and the third is the base of the cylinder. (This arrangement is pretty ingenious, in that it removes the need for a "cone" command; change the value of either the second or third parameter to zero and see what you get.)

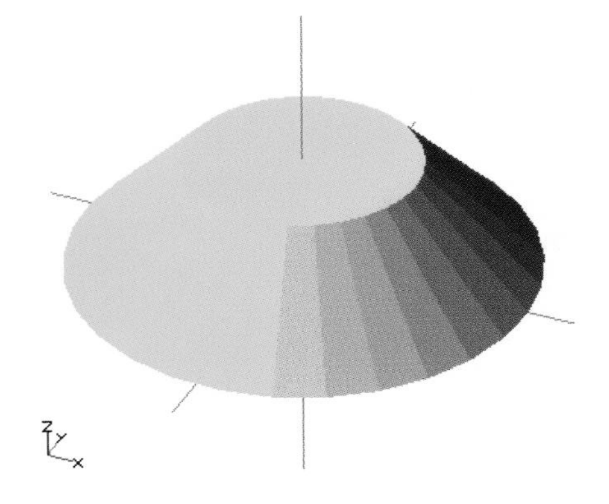

Figure 11-11 *The cylinder function can create cones and truncated cones*

If you want a pure cylinder, instead of using two radius parameters, `r1` and `r2`, you can use only one, allowing the other radius to default to the first. For the cylinder, as well as for the sphere, you have the `$fn` parameter to determine the resolution.

Variables

In OpenSCAD you can use variables to store information you are interested in, like numbers or words, which you can use later. Each of these variables has a name to distinguish it from the others. For example, you can name a variable `side`, and use it to represent, say, the side of a cube. You can assign a numeric value to `side` like this:

```
side = 7;
```

In this way you could create a cube with the following instruction:

```
cube(size = side);
```

You can use this variable for any shape:

```
circle(size);
```

Or even:

```
cylinder(size, size);
```

Why not just use the number 7 for all these shapes instead of a variable? Because if you want to change the size of all the objects at once, you only need to change the value of `size`.

Move Slightly!

To create a second cube, you have to change some parameters. For example, type these three instructions into OpenSCAD:

```
side = 5;
cube(size = side);
cube(size = side);
```

Press F6 and see what happens. It looks like it hasn't worked, because you can see only one cube. Actually there are two, but they are identical and in exactly the same place (the physical problem of two identical objects occupying the same intrinsic space at the same time doesn't apply here). What can you do?

You can create an object in any location of the design space by using the `translate` function, specifying how far from the origin you want to move it along the three axes. Replace the last line of the code with the following two lines:

```
translate([10,0,0])
cube(5, center = true);
```

In this example, we have centered the cube on the origin and then moved it by 10 mm, as shown in Figure 11-12. You must not put a semicolon after the `translate` instruction, because the semicolon would terminate the instruction, and you need `translate` to apply to the next object.

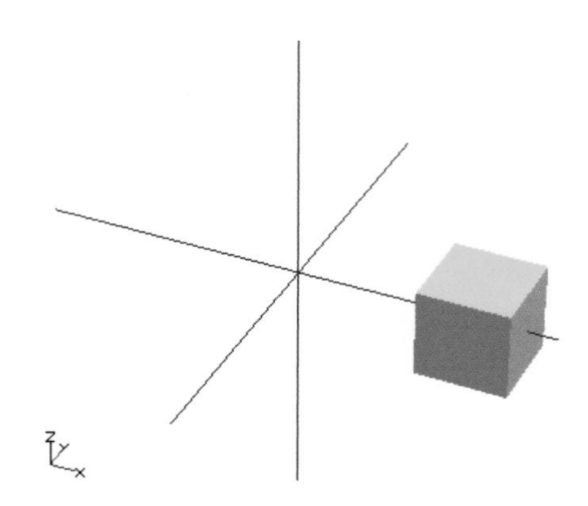

Figure 11-12 *A cube with a 5 mm side shifted along the x-axis and centered*

Now we've got it! But it's not over yet...

Should you want to create a row of identical cubes, all at the same distance from one another, you could simply write an instruction for each cube, specifying how far you want to move it from the origin (you don't have to move the first cube):

```
cube(size = 1, center = true);
translate([0, 0, 2])
  cube(size = 1, center = true);
translate([0, 0, 4])
  cube(size = 1, center = true);
translate([0, 0, 6])
  cube(size = 1, center = true);
translate([0, 0, 8])
  cube(size = 1, center = true);
translate([0, 0, 10])
  cube(size = 1, center = true);
```

You have achieved the goal, but the situation risks getting out of hand very quickly.

Lazy Is Good!

We are fortunate that OpenSCAD doesn't provide us only with simple instructions, but also with a real programming language that is perfect for such repetitive commands. We want to draw a cube with a 1 mm side, then move upward by 2 mm and draw an identical cube, again and again, until we get to 10 mm from the origin.

Let's translate our intention in the OpenSCAD language. To repeat one or more operations for an arbitrary number of times, we will use a structure called a *loop*.

The set of rules to create a loop is simple:

```
for ( z = [0 : 2 : 10] )
{
  translate([0, 0, z])
  cube(size = 1, center = true);
}
```

Essentially, you are saying to OpenSCAD: "Do what is included in the curly braces. Start with z equal to 0, and then repeat, increasing the value of z by 2 until you reach 10." You could also say that, with the first line you create a sequence of even numbers from 0 to 10, and that at each iteration you repeat the block in curly braces, replacing the z variable with its current value. Figure 11-13 shows the result.

The result is the same, but the script is much more compact and, after you have some experience, even more simple to read: in the prior example you place some cubes in predefined positions, but in this version, you can immediately understand that the origins of each cube are placed at the distance of 2 mm from one another. In the first listing the information is hidden, while in the second it is evident.

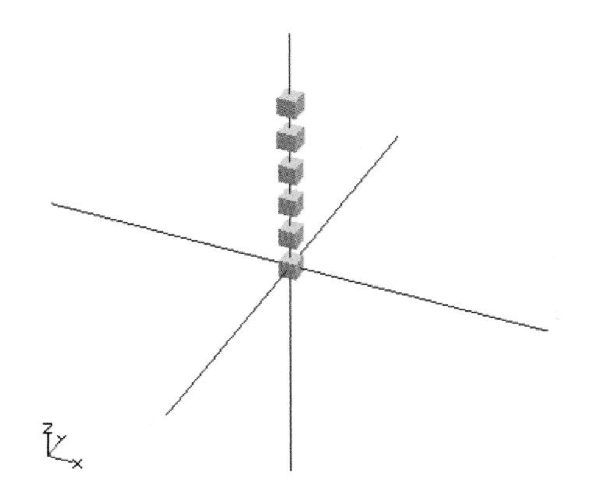

Figure 11-13 *Six cubes in a row*

You can also specify a precise irregular range of values:

```
for (z = [-2, 2, 6])
{
  translate([0, 0, z])
  cube(size = 1, center = false);
}
```

What's the difference between this and the previous listing? The earlier listing, which had colons separating the entries in the `for` loop, iterates over a *range* of values. In that case, we told the software to count from 0 (the first value) to 10 (the last value), incrementing by 2 (the middle value). The software then figures out the values (0,2,4,6,8,10) it should use.

When you separate the values in a `for` loop with commas, we tell the software to use *those exact* values, and only those values. In the second example, we tell the software to use the values –2, 2, and 6—and no others.

You can also use the `translate` function to move more than one object, grouping the necessary instructions in brackets:

```
translate ([5,0,0]) {
  cube (1);
  translate ([0,3,0])
    sphere (1);
}
```

Note how, after having moved along the x-axis in the first instruction, before generating the sphere, we move again along the y-axis: you are *nesting* the two translation movements, one inside the other (Figure 11-14).

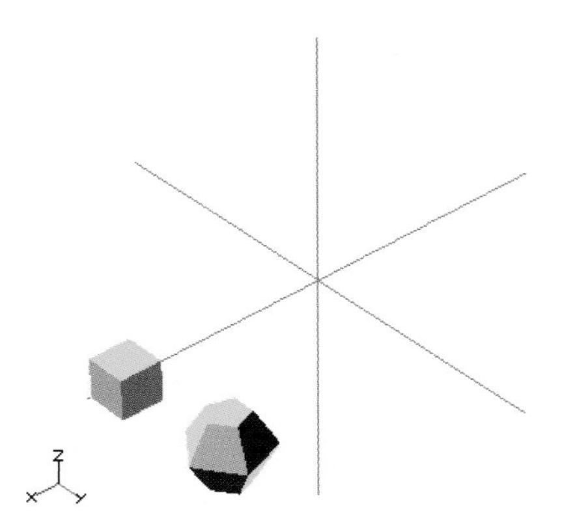

Figure 11-14 *Shifting both a cube and a sphere*

Other Transformations

The `scale` function resizes an object along the three axes:

```
cube(10);
translate([15,0,0])
  scale([0.5,1,2])
    cube(10);
```

In that example, we have drawn a cube with a side of 10 mm; then we have taken a similar cube and moved it on the x-axis by 15 mm, after applying the `scale[0.5, 1, 2]` function to it. The result is to halve the size of the cube side along the x-axis, leave the dimensions unaltered along the y-axis, and double the dimensions along the z-axis (Figure 11-15).

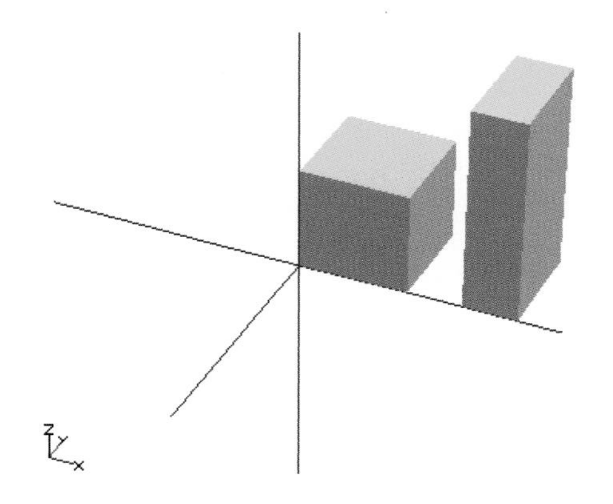

Figure 11-15 *Modifying the objects' look with the scale function*

Sure, this function doesn't seem very useful on a cube, but you can use it on a sphere to obtain a rugby ball or a flying disk.

To rotate the objects, you use the `rotate` function. To see how it works, draw a cone with the basis centered in the origin, then make a second cone identical to the first one and apply to it a 180-degree rotation around the y-axis, so that it turns upside down (Figure 11-16):

```
cylinder(10,7,0);
translate([10,0,10])
  rotate(a=[0,180,0])
    cylinder(10,7,0);
```

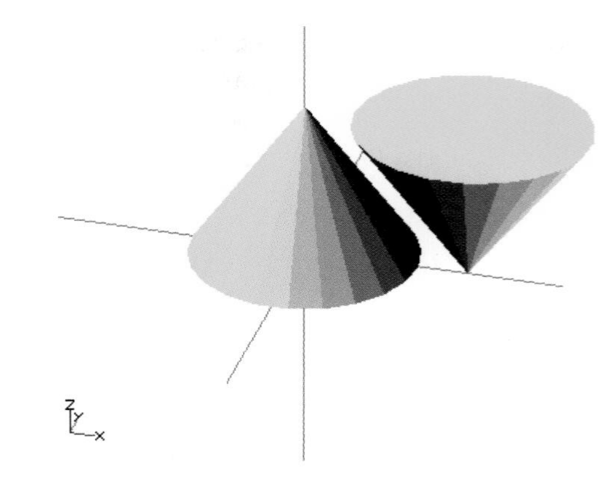

Figure 11-16 *Objects and rotations*

The optional `v` parameter specifies an arbitrary axis to rotate around:

```
cylinder(10,7,0);
translate([10,0,10])
  rotate(a=60, v=[1,1,0])
    cylinder(10,7,0);
```

Practically, we define the axis around which we want our shape to rotate with a vector whose components along the three axes are `[1, 1, 0]`; stated more simply, the direction of the axis is the direction of the segment that links the origin with the point of the (1,1,0) coordinates. The `a` parameter indicates by how many degrees you want your object to rotate around this axis (Figure 11-17).

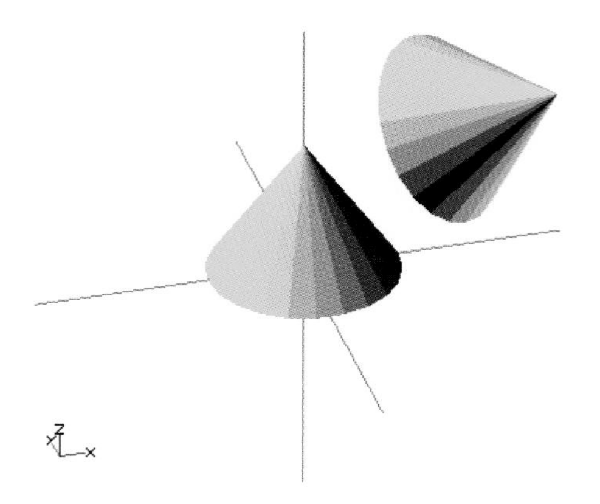

Figure 11-17 *Rotation around an arbitrary axis*

The `union` command unifies two or more objects, combining them into one; here's an example from OpenSCAD's online manual:

```
union() {
  cylinder (h = 4, r=1, center = true,
$fn=100);
  rotate ([90,0,0])
    cylinder (h = 4, r=0.9, center = true,
$fn=100);
}
```

Let's look at what it does. It creates a cylinder, then creates another, slightly smaller cylinder at a 90-degree angle to the first one. Both cylinders are centered at the origin. The result is shown in Figure 11-18.

Just as it is possible to unify objects, you can also create a `difference`, which subtracts one object from the other. Let's replace `union` in the previous example with `difference`:

```
difference() {
  cylinder (h = 4, r=1, center = true,
$fn=100);
  rotate ([90,0,0])
    cylinder (h = 4, r=0.9, center = true,
$fn=100);
}
```

See the result? Much as in the `union` example, the `difference` command creates a cylinder, and then creates another, slightly smaller cylinder at a 90-degree angle to the first one. Only this time, instead of combining the two, `difference` *subtracts* the second object from the first.

OpenSCAD offers functions for two-dimensional plane figures, too. The `square` command draws squares and rectangles; just like the `cube` function, it accepts the `center = true` parameter:

```
square ([2,2], center = true);
```

To draw a circle we can use `circle`. For `circle` you can again indicate the resolution with the `$fn` parameter:

```
circle(2, $fn=50);
```

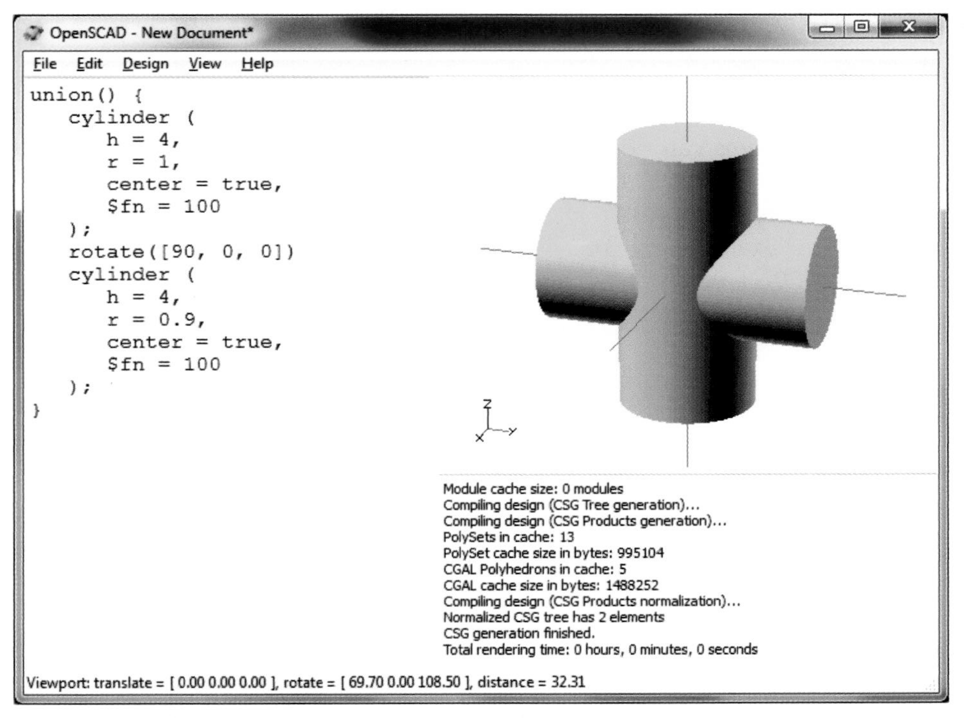

Figure 11-18 *Union of two cylinders*

There is no command to draw ellipses, but it is always possible to deform a circle. Here is a completely flat ellipse, with a 40 mm main axis and a 20 mm secondary axis (Figure 11-19):

```
scale([2,1,0])
  circle(20);
```

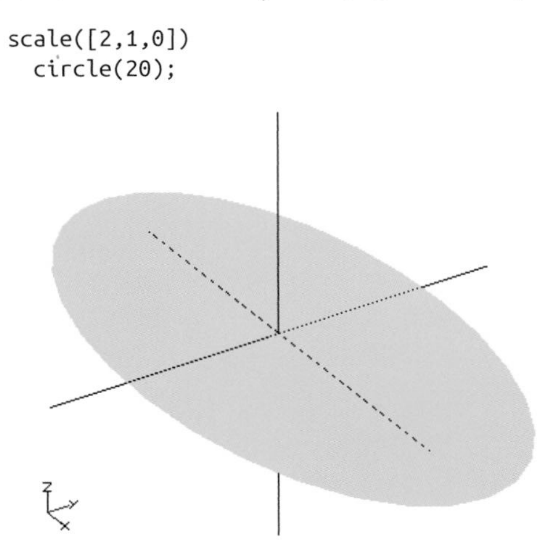

Figure 11-19 *We obtain the ellipse by deforming a circle*

You can also draw polygons by indicating a series of points and the paths linking them:

```
polygon(points=[[0,0],[100,0],[0,100],
[10,10],
[80,10],[10,80]],
        paths=[[0,1,2],[3,4,5]]);
```

The two-dimensional figures can be easily extruded into 3D objects with the `linear_extrude` command. Let's try to extrude the polygon of the previous example by 100 mm, shown in Figure 11-20. If you don't extrude your 2D objects, they won't be printable:

```
linear_extrude(height = 100)
  polygon( points= [
      [ 0, 0], [100, 0], [ 0,100],
      [10,10], [80,10], [10, 80] ],
    paths=[ [0,1,2], [3,4,5] ]
);
```

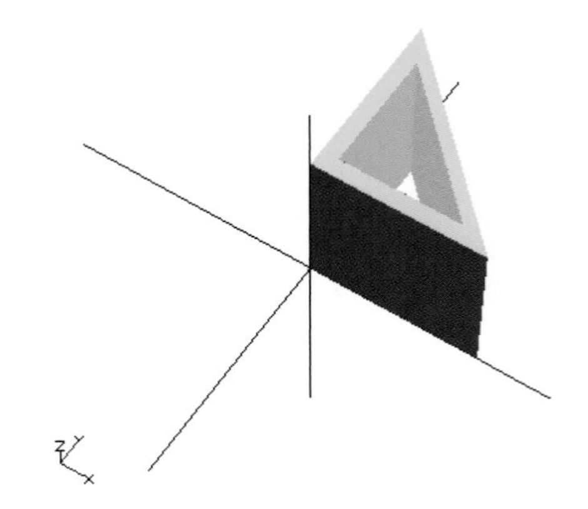

Figure 11-20 *Extruding a polygon*

The `linear_extrude` command accepts other parameters too; for example, the `center` option is used to center the object while it is being extruded:

```
linear_extrude(height = 20, center = true)
  circle(r = 10);
```

OpenSCAD also has a `rotate_extrude` function that rotates the object around the z-axis, allowing you, for example, to create toroidal (donut-shaped) objects (Figure 11-21):

```
rotate_extrude()
  translate([20,0,0])
  circle(r = 10);
```

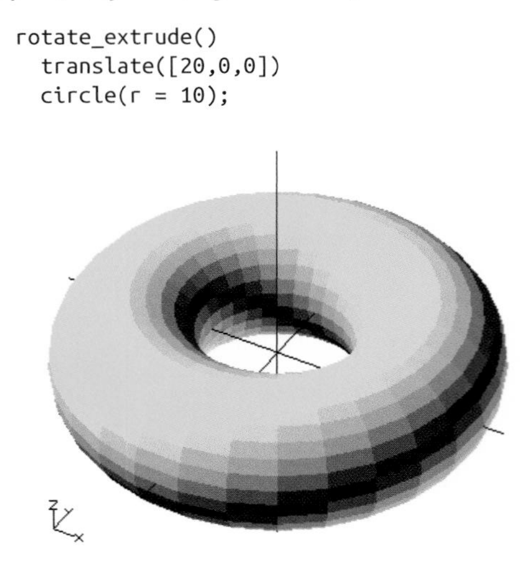

Figure 11-21 *A toroid*

For further details on the extrusion functions, see the OpenSCAD online manual (*http://www.openscad.org/cheatsheet/index.html*).

Expanding OpenSCAD

Unfortunately there is not a native method to create text. But, luckily, you can expand OpenSCAD by using *libraries*, which are software packages that provide additional behaviors not included in the original version of the software.

On Thingiverse there is a library created by Phil Greenland (*http://www.thingiverse.com/thing:59817*) that allows you to turn text into three-dimensional objects. Download the *TextGenerator.scad* file. To make a library available for all your projects you have to copy it into a specific folder, depending on the operating system you are using; usually the directories are:

- Windows: *Programs\OpenSCAD\libraries*
- Linux: *$HOME/.local/share/OpenSCAD/libraries*
- OS X: *$HOME/Documents/OpenSCAD/libraries*

Alternatively, to use a library on a specific project only, you can copy it into the folder where you saved your OpenSCAD project file.

To use the library in a script, you have to load it with the `use` command. You can then write the text you want with the `drawtext` command (Figure 11-22):

```
use <TextGenerator.scad>
drawtext("Hello World");
```

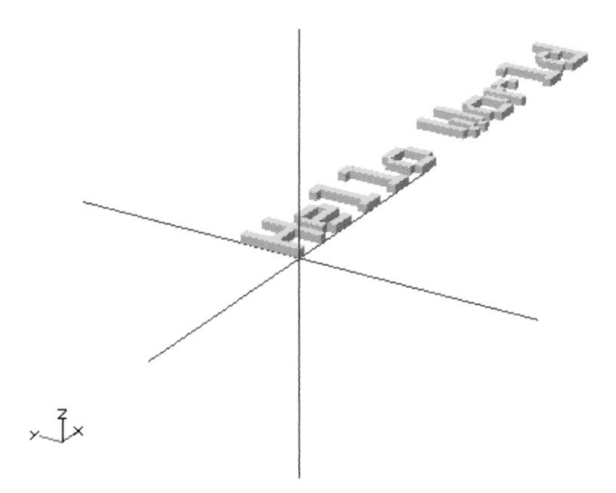

Figure 11-22 *Hello World*

3D Printing

12

Noted science fiction author Arthur C. Clarke once said: "Any sufficiently advanced technology is indistinguishable from magic." If you look at a 3D printer at work, you can understand what he meant (see Figure 12-1). Seeing an object taking shape layer by layer, or emerging from a resin bath for the first time, is like witnessing magic. These technologies are entering our homes, and they are accessible at reasonable cost. With a device that costs approximately as much as a mid- to high-end laser printer—around $600 to $2,500 at the time of this writing—you can create nearly any kind of object you like, or fix broken objects that nobody sells or manufactures anymore. Welcome to the world of self-production!

3D printers can be simple to operate. Most recent printers require nothing more than spooling up the source material (usually some form of thermoplastic), inserting an SD card that contains the design file, and pushing a button. The printer reads the design file, melts the source material, and layers it until the desired object is completed.

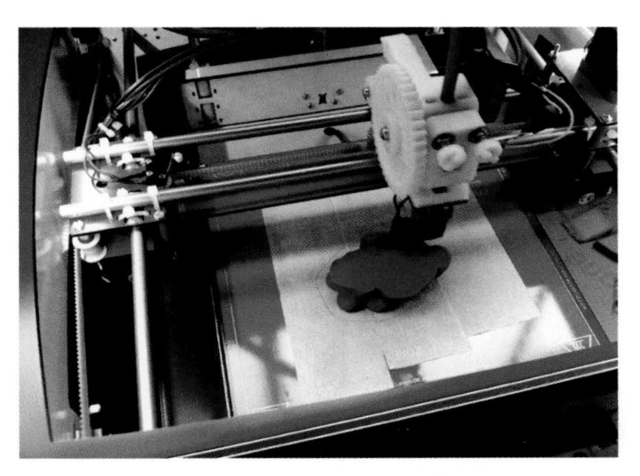

Figure 12-1 *Printing an object, layer by layer, on a Sharebot Kiwi*

This kind of additive technology is called *fused deposition modeling* (FDM) and is, at the moment, the most accessible and cost-effective option, so it's the one we'll be talking about. Other technologies involve sintering (hardening of powders) by means of a laser, stereolithography (polymerization of resins by exposure to a light source), and layering of shaped plates (laser cut or milled).

How Does It Work?

An FDM printer builds the object on a platform that moves vertically, on its z-axis. An *extruder* —a device with a nozzle and a heating element, very much like a hot glue gun—moves on the x- and y-axes and melts a plastic material, placing it layer by layer on the platform (see Figure 12-2). Thanks to the high-quality *stepper motors* used, many FDM printers can reach a precision of movement of a few tens of microns on the x- and y-axes, and 2.5 microns (about 0.0001″) on the z-axis. However, because of the extruder dimensions, the resolution is generally between 100 and 350 microns (0.004″ to 0.014″) per printed layer. There is at least one motor for each axis and one for the extruder, even though in some cases two motors can be placed on one axis. The circuits that direct the steppers are called *drivers*.

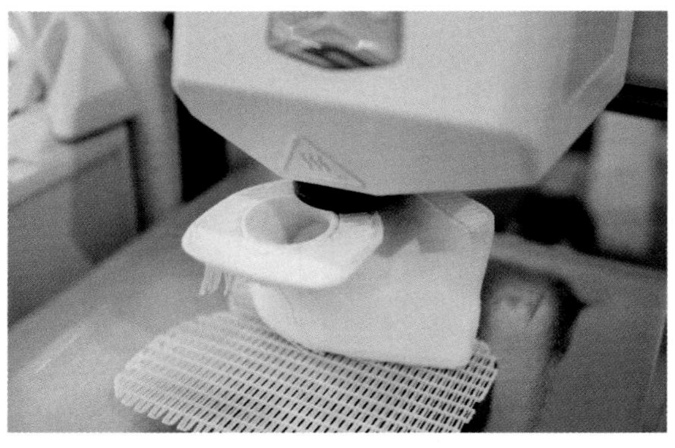

Figure 12-2 *An extruder at work at the FabCafe in Tokyo*

The extrusion nozzle, a kind of metallic funnel with a tiny hole, is connected to a resistor that heats up to around 250°C (some types of extruders can reach temperatures of over 300°C). The extruder is heated up to a higher temperature than the melting point of the materials used, so that the materials reach a perfect flow and can be layered on the platform without dripping too much.

The size and structure of the printer define the printing area. In the early models, the working area was a few centimeters across; today the most widely used area is 20×20×20 cm, because that is the best compromise for a desktop printer, even though some models can work on a larger area. While this printing area might seem small, you can easily create larger objects by building them in pieces and assembling them later. Some printers have minimal structure, with no frills, making them inexpensive and easy to assemble. Other printers have an external structure that makes them look like vintage televisions or microwave ovens.

Materials

The two most common printing materials, which come in spools of filament of different thickness and colors, are *acrylonitrile-butadiene-styrene* (ABS) and *polylactic acid* (PLA). They look similar, but they have very different properties. For supporting structures, *polyvinyl alcohol* (PVA) is also used.

ABS is a thermoplastic material, easy to get and widely used in injection molding. Even if you haven't heard the name before, it is famous for being the material LEGO building blocks are made of. It melts at 105°C, though in 3D printers it is extruded between 215°C and 250°C. Even though it smells like burnt plastic during printing, it is not toxic. However, while the actual plastic itself is safe to handle, the vapors it gives off when heated can be irritating, so most makers will use ABS only in a well-ventilated room, or one with a fume extraction system.

ABS also has a tendency to deform as it cools down at different rates in different printed areas; to avoid this you could use a kind of oven that keeps the whole object at a steady temperature, but this technology is still covered by patent.

What's the Use?

Nowadays, desktop 3D printers give makers many opportunities in the most diverse fields:

- They can print robot parts, as they do, for instance, at the Robotics and Artificial Intelligence Laboratory of the Politecnico of Milan, or in the workshop of robot maker Michael Overstreet (*http://bit.ly/1NyOzqR*).
- They can fix objects by producing parts that don't come as spare parts, saving you the purchase of a brand new object: for instance, a joint for a bookshelf, the detergent compartment of a washing machine, etc.
- They can make customized objects, such as universal joints for a laser cut steadycams, mobile phone cases, electric guitar bodies, any size, shape and feature and boxes of any size, shape, and feature for electronic projects.
- During product development they provide a fast iteration cycle: a maker designs a part, prints it, tests it in the available parts, and if it's not quite as she likes it, the computer file can be easily changed and printed again, repeating the cycle.
- They can make replicas of no-longer-marketed objects, like toys from your childhood.
- In the hobby crafts field, they can create toys, models, soldiers, and landscape details.
- They can print small busts portraying your friends: at Disney it is even possible to buy a stormtrooper with your face or get frozen in carbonite!
- In architecture, they can create building models; some people are even considering the possibility of making houses.
- They can create artistic objects.

PLA is a natural, biodegradable thermoplastic material derived from corn or potato starch; it is extruded between 160°C and 220°C and, once cold, it is slightly more rigid than ABS, but more fragile. It can be printed on a nonheated platform and does not pose any deformation issue while cooling down, as ABS does. However, the quality of the printed objects is slightly lower. PLA also suffers from its own virtue: if it is used or stored in conditions of over 90% humidity, or a temperature over 60°C, its biodegradation process starts, and it slowly turns into mush.

PVA is easy to use for printing because it melts at a lower temperature than ABS and PLA, and doesn't need a heated platform. It is much easier to remove since it dissolves in water. Its filament poses the same storage issues as PLA.

For information on new and interesting materials for 3D printing, be sure to check the archives of Matt Stulz's posts for makezine.com (*http://bit.ly/1NyOQtP*), many of which feature reviews of various 3D printing materials.

3D Printers

In the industrial world, 3D printing technology goes back to the 1980s. How did it make it to the desktops of the 2010s?

An early pioneer of desktop 3D printing was Dr. Adrian Bowyer, a mechanical engineering professor at the University of Bath in England. In 2005, Bowyer started prototyping a low-cost, open source machine, the RepRap (*http://reprap.org*), which stands for *replicating rapid prototyper* (see Figure 12-3). The eventual goal was to have the RepRap make all the necessary plastic parts to assemble its own replica. The machine has a simple structure, made out of threaded bars and easily available components. The first prototypes were sold in kits.

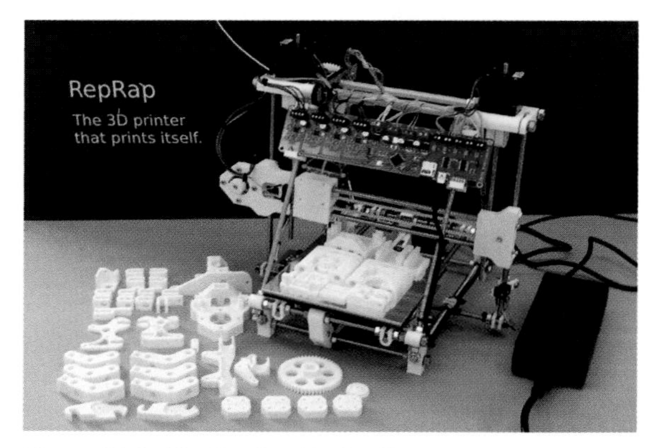

Figure 12-3 *The RepRap Huxley 3D printer by RepRapPro Ltd (http://reprappro.com)*

Later, with the RepRap project as a starting point and with a few changes to its structure and mechanisms, many people designed and made their own version of a 3D printer. We'll discuss just a few.

MakerBot

In 2009, some time after the appearance of the RepRap project, Bre Pettis, Adam Mayer, and Zach "Hoeken" Smith founded a company called MakerBot (*http://www.makerbot.com/*), manufacturing open source 3D printers. Cupcake, their first model, was a great success; later, Replicator and Replicator2 came along. MakerBot no longer makes open source 3D printers.

Kentstrapper

In Florence, Italy, three generations of the Cantini family have given life to an established startup named Kentstrapper (*http://www.kentstrapper.com*), where they develop, manufacture, and sell printers derived from the RepRap model. They have been the pioneers of 3D in Italy. Their most popular models are the Kentstrapper Volta, Mendel Max (shown in Figure 12-4), and RepRap Galileo.

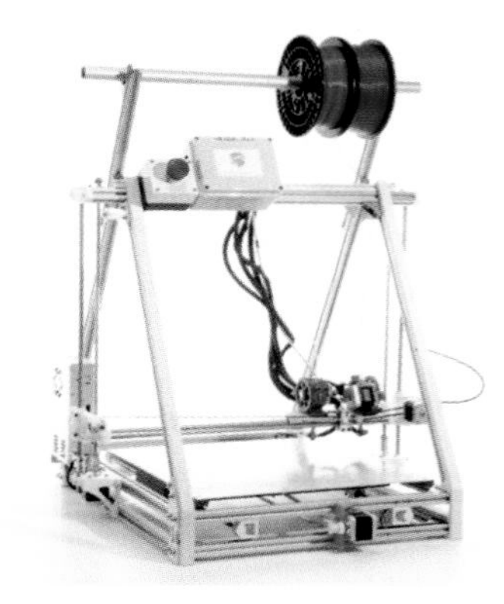

Figure 12-4 *A Mendel Max with an oversized platform for large printings (Kentstrapper)*

WASP

Massimo Moretti, fascinated by the creations of thread-waisted wasps and with the help of a group of students at the ISIA Design University, started WASP (*http://www.waspproject.it*) (World's Advanced Saving Project) and produced PowerWASP, a desktop *personal factory* that can print plastic and clay and carry out light milling work (see Figure 12-5). Massimo's goal is to one day build a machine able to print clay dwellings, in the way wasps do.

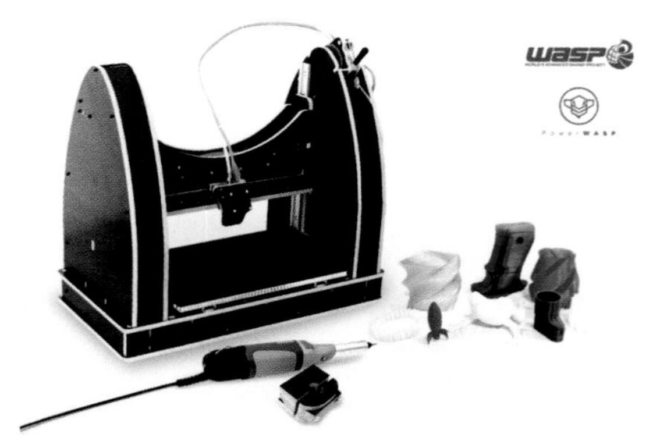

Figure 12-5 *The PowerWASP printer can also print clay and do milling work (WASP)*

Many 3D printer manufacturers offer their printers as kits. This way, the price can be kept reasonable and you get the pleasure of assembling a complex object with your own hands. The task might require a few days and some manual work, but it is worth it, because in the end it will be "your" printer, and you will be familiar with every single screw. Printrbot and Ultimaker both make printers that are very popular with makers; some of their models are available as kits, such as Printrbot's Simple Maker's Kit (*http://bit.ly/1NyQbRI*), and Ultimaker's Ultimaker Original+ (*http://bit.ly/1NyQcVu*).

There are more printers listed in the RepRap Family Tree (*http://bit.ly/1NyQjAp*), and once a year, Make: magazine devotes an issue to reviews of the latest and greatest 3D printers (*http://makezine.com/3d-printing/*).

The Workflow

The first step to print a 3D object is to design, or obtain, a 3D model. You can create it with one of the software programs you got to know in Chapter 11, or search online to find out if anyone has already created what you need. Two places to start are Thingiverse.com (Figure 12-6) and YouMagine.com, websites that contain many open source 3D models for you to download, modify, and print. Starting from a model someone else has created is a great idea, because you can learn from people who are more experienced with designing a model.

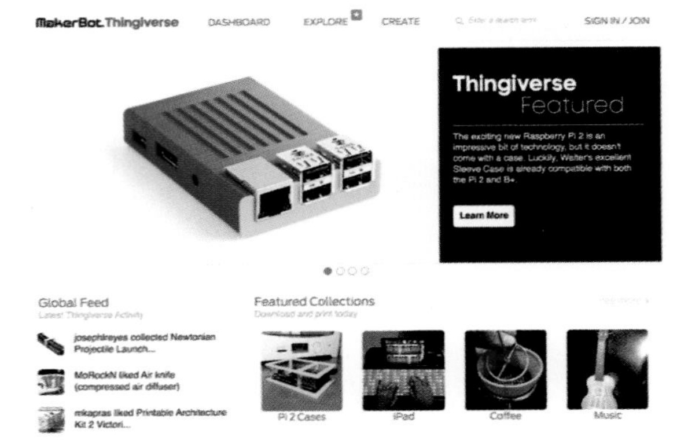

Figure 12-6 *Thingiverse*

In the construction of movable parts, you mustn't overlook *tolerances*, or the ability to cope with imperfections. Even precision gear manufacturers consider tolerances and leave room for movement, and you have to as well. Trying to produce all the parts of your moving object with pinpoint accuracy is not a great idea, because they will hardly ever fit together as gracefully as they do on your computer screen. You have to keep in mind the mechanical features, the printing area, and the resolution of the machine that is going to produce the object: metaphorically speaking, there is no point in designing finely chiseled details when the only equipment you have is a chainsaw.

If you wish to reproduce something that already exists, you can try to 3D scan it and then print it. You can use a dedicated 3D scanner, or you can look at some of the 3D scanning apps for the Microsoft Kinect or Structure Scanner (*http://structure.io*), or you can take a sequence of pictures from many different angles and use an application like *Autodesk 123D Catch* to use the pictures to reconstruct a model of the object. These techniques require some expertise, because scan-generated files often contain defects that need to be rectified with specific programs; still, this method can be quicker than creating a brand new model.

Corrections

Sometimes your computer models look perfect because you are observing and handling three-dimensional objects with two-dimensional screens and tools. In fact, such seemingly "perfect" models may actually have so many serious defects as to make printing impossible.

Perpendiculars to the plane

We said that models are approximated through polygons; for the STL format, for instance, a triangle mesh is used. Each polygon for the model has an internal and an external face. You can represent the external face with an arrow perpendicular to the surface or, to be more precise, with a vector that is *normal* to the plane, as shown in Figure 12-7.

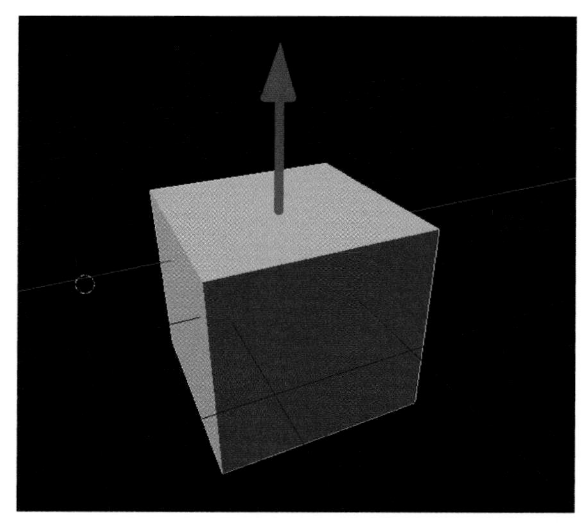

Figure 12-7 *The normal to a cube plane*

In a correct model, all normals have to point outside. Even though this doesn't seem very relevant, it is important for a printing program, because it needs to understand what the internal and external parts of the object are.

Watertight models

Another issue is the presence of gaps and unclosed surfaces: all models need to be *watertight* (see Figure 12-8). This problem often arises when you create a model starting from a 3D scan.

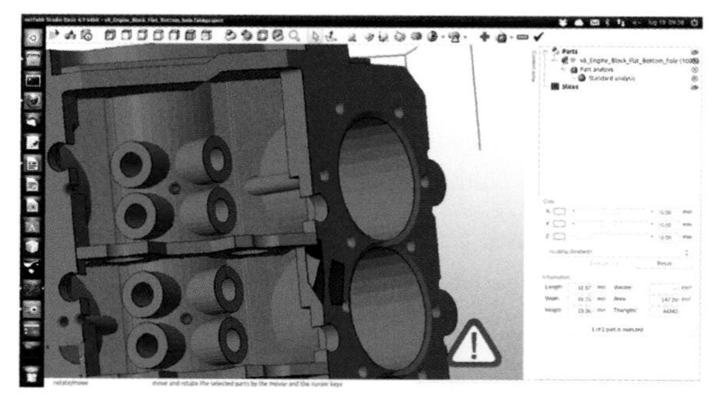

Figure 12-8 *In this model, the red surface is not "watertight"*

Before printing, you can use software to check whether everything is in order. These programs are basically viewers with simple editing capabilities, like resizing objects or moving them to an ideal position for printing. They are not as complex as a CAD program, even though some can identify and amend the most frequent mistakes that models can present, such as closing gaps and merging unconnected parts.

Netfabb Studio Basic (*http://bit.ly/1NyS8x9*) (Figure 12-9) is a free program that displays, analyzes, corrects, and edits STL files on Windows, Linux, and Mac. It can detect models that are not watertight.

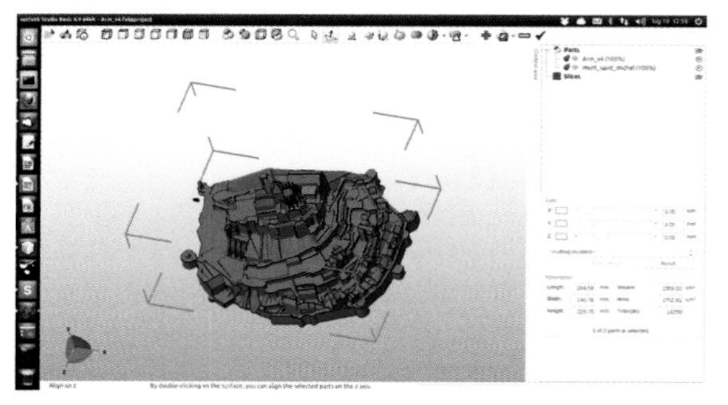

Figure 12-9 *Netfabb Studio Basic*

Pleasant3D (*http://www.pleasantsoftware.com/developer/pleasant3d/*) (Figure 12-10) is available only for Mac and allows users to make small changes to files—rotating, resizing, centering—in order to obtain the best possible printing of the file. It can also read the format that is actually used for printing, thus allowing you to see how the extruder is going to place the material, layer by layer.

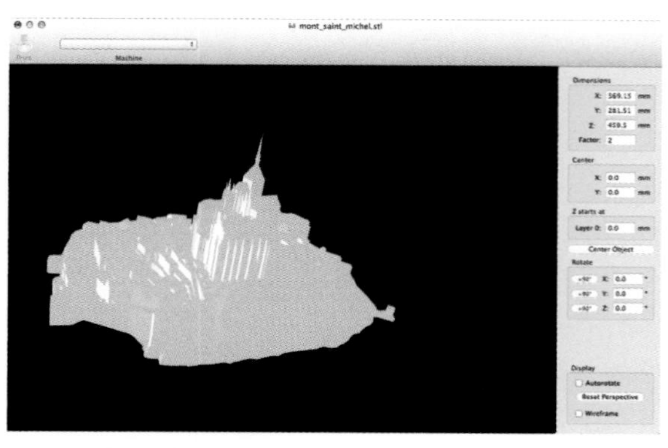

Figure 12-10 *Pleasant3D*

MeshMixer (*http://www.meshmixer.com*) (Figure 12-11) is freeware, but not open source software. Created by Autodesk, it can modify STL files and create 3D models. It can also detect nonwatertight files.

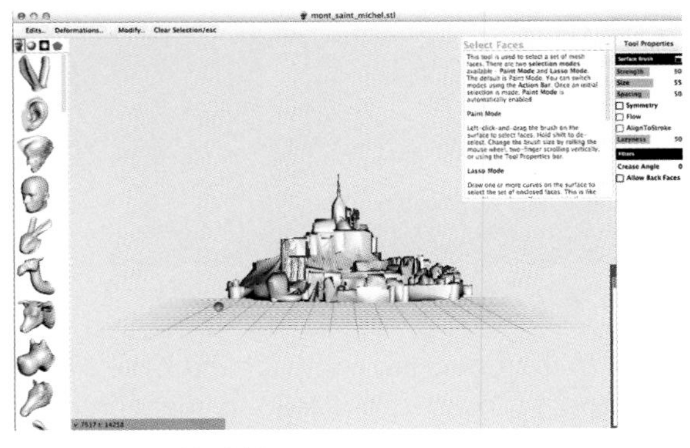

Figure 12-11 *MeshMixer*

MeshLab (*http://meshlab.sourceforge.net*) (Figure 12-12) is another open source package to visualize and modify STL files, available for Windows, Linux, and Mac.

Figure 12-12 *MeshLab*

Slice It Up!

Although one of the most common formats for 3D models is STL, 3D printers cannot use this format directly, so you have to first convert STL into a sequence of commands that coordinate movement and operations. The standard language for this is called *G-code*.

In the G-code conversion, you have to consider the fact that the extruder will have to layer the material, so basically you have to "slice" your model horizontally; in fact, the term *slicing* is often used. Again, there are software programs that can help here; one of the most popular is *Slic3r* (*http://slic3r.org/*) (Figure 12-13), an open source program by Alessandro Ranellucci. It is a simple program, it is easy to use, and it can manage complex tasks; it makes use of many mathematical algorithms to assess the best paths to produce the object. Among other things, Slic3r can also evaluate the amount of material to be extruded and check temperature, fans, and cooldown time.

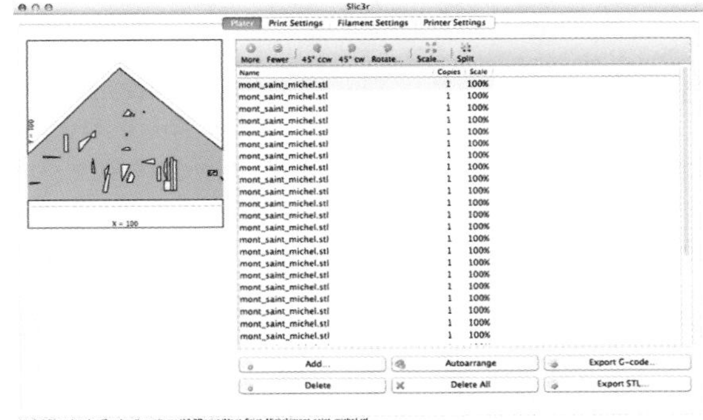

Figure 12-13 *The Slic3r software at work*

Setting Up the Printer

A 3D printer is an object with moving parts that are not always protected, and one or two extruders that reach very high temperatures, so it is wise to place it on a secure and stable surface such as a table or a desk, away from pets and small children. Check carefully the positioning of the power cable, of the USB cable for the computer connection, if any, and of the plastic filament spool.

Adjusting

The first thing you need to do before printing is to adjust the printing platform, which needs to be at the right distance from the nozzle: not too close and not too far. If it is too distant, the extruded thread won't be laid in place; if it's too close, it might clog the extruder. With the current generation of printers, the aligning procedure is very easy: just select the relevant feature and follow the instruction on the display. In older models, the procedure is manual and you have to move the extruder along the platform and check the distance by hand.

It is recommended to follow this procedure also for preassembled printers just out of the box, because if they're jostled in transit it might alter their factory settings.

Extruder feeding

From the PLA or ABS spool, examine the free end of the filament line: it needs to have a clean cut; otherwise, it might get stuck or not fit in properly. Check that the line and the runner tube, if any, are well positioned; the spool must be free to spin. Slide the line through the runner tube and into the extruder. The extruder must have reached the right temperature; be careful, since it will be over 180°C.

Older extruders were made of 3D printed parts and included some sort of trigger that pressed the filament against a little knurled wheel, which forced the filament into the nozzle. These models were open and it was possible to see where the filament went as well as any possible obstruction. Figure 12-14 shows a close-up of such an extruder.

Figure 12-14 *The extruder of a Sharebot Kiwi*

Modern extruders are compact and have a little opening to insert the plastic filament into, as shown in Figure 12-15. The filament must be pushed in firmly and as soon as a thin trickle of plastic starts to drip from the nozzle, the extruders are ready to print.

Figure 12-15 *A more modern extruder*

Operating the Printer

If you don't print from an SD card, you can operate the printer from a computer that can execute a program to send out manual input, receive diagnostic information, check on the work and temperatures, and also estimate time and layers that will be produced. The most popular programs are Printrun (*http://www.pronterface.com*) (Figure 12-16) and Repetier-Host (*http://www.repetier.com*) (Figure 12-17).

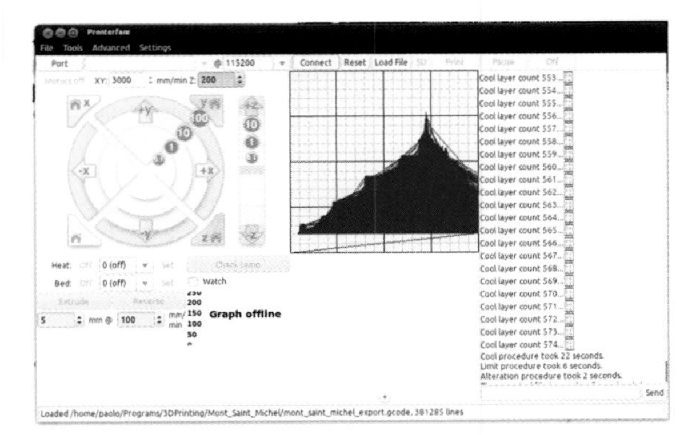

Figure 12-16 *Printrun*

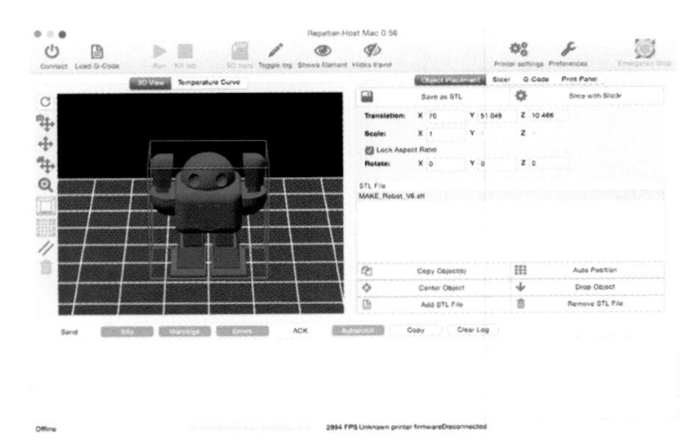

Figure 12-17 *Repetier-Host*

An SD card comes in very handy when you want to print an object that you have already tested and for which you have already set up all the printing parameters, or if you don't feel like carrying your laptop along, or if you have more than a couple printers working in parallel. In all other cases, operating the printer through a computer is a fine choice.

Now let's cover some of the options you can configure in your slicing software.

Shell

Each layer of the object you are printing is made up of an outer *shell* as well as interior *infill*, described next. You can change the thickness of this shell by specifying the number of perimeters: the perimeters are concentric and are added inward to keep the external dimensions of the object consistent, so the more perimeters (3 is a good choice), the less infill you'll get. A thicker shell makes the object more rigid.

Infill

Most 3D printed models are mostly hollow. They're not usually completely hollow, because they need some sort of internal structural support to keep from falling apart. The infill is the extent to which the interior of the object is filled. It is usually determined by the algorithms used for the model slicing. Its typical value is 10%, meaning that everything inside the shell will be 10% material and 90% hollow. An object with 100% infill will be heavy and compact, with no internal empty space; it will require a higher amount of material and a long printing time. A high infill value makes the objects stronger and more resistant to damage. Your slicing software may allow you to choose different infill patterns, from rectangles to hexagons, so you should experiment to find out what gives you the strongest object.

Bridging

Printers can print objects with an *overhang* (sections of the print that hang over empty space), though only with a maximum angle of 45 degrees. Some of the solutions to printing over empty space are managed directly by *bridging* algorithms, which take care of the necessary adjustments. The final outcome will usually show some frayed edges (Figure 12-18). It is also possible to create structures with sharper angles or with bridging parts, though in these cases it is necessary to create supporting structures, such as very thin and regular walls, and include them in the model. In both cases, at the end of the printing process you will get rid of the supporting structures. You could also enable support structures, but these will need to be removed and may leave unsightly marks on your object.

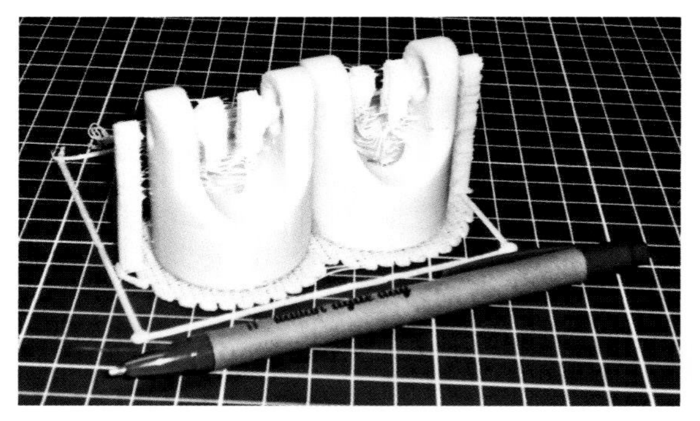

Figure 12-18 *Two elements of a universal joint with suspended parts: note the inevitable "smudges" around the suspended parts*

If you have a dual extruder, you can use different materials for the supporting structures, such as polyvinyl alcohol (PVA), which dissolves in water.

Raft

To prevent corner warping problems, you can place a thin layer, called a *raft*, on the platform and print your object on top of it. The raft is also helpful to improve bonding with the printing platform. Another technique used to improve bonding is to use a glass platform sprayed with hairspray or covered with a glue stick. Never apply hairspray directly on the printer, because the gears might seize up. Another technique used to prevent the object from deforming during the printing process is to add structures called *shields*, *baffles*, or *mouse ears* in appropriate positions.

Finishing off a printed object

Sometimes printed parts show an unwanted surface texture. You can refine the objects with sandpaper, varnishes, or other finishing, to make the surface perfect. If you use sandpaper, it is important to avoid inhaling the resulting filings. For the best result start with coarse paper and proceed toward a finer and finer one; even though it might seem more efficient, starting with fine paper will never lead to good outcomes.

Finally, you can use a heat gun to heat up the surface slightly and make it glossier, as cake designers do with chocolate frosting.

Figure 12-19 *Production of a complex object with supporting material (Matt Stultz, 3dppvd.org)*

What If You Don't Have a Printer?

If you don't have a 3D printer yet, don't despair: you can make use of one of the many online prototyping services. Some popular ones are Ponoko (*https://www.ponoko.com*), 3D Hubs (*https://www.3dhubs.com*), i.materialise (*http://i.materialise.com*), Shapeways (*http://www.shapeways.com*), and Sculpteo (*http://www.sculpteo.com/en/*).

After creating an account, simply upload a file and choose a material from their extensive catalogues, which may include some metals depending on where you order your print. In a few clicks you can finalize your purchase, and after a few days or weeks your high-definition creation will be shipped to your doorstep. The cost basically depends on the chosen material, size, and filling parameters. Pretty much all services offer advice for the creation of the models you wish to print; some will even point out errors and problems in your model, if any.

Milling 13

Michelangelo, artist and master of subtractive techniques, used to say that what he did was to free his works from the marble that kept them trapped. Today, thanks to machines, we can also free "objects" trapped in foam blocks, resin, or wood, without necessarily becoming artists of the chisel. These machines, which make our objects rise from an ocean of burrs, are called *computer numerical control* (CNC) machines—that is, machines that can control other tools (such as mills, lathes, grinders, and drills) via computer.

CNC Machines

Just like 3D printers, CNC machines are nothing new: they have been used for decades to manufacture precision mechanical parts. These machines are often very large and expensive, made of cast iron and filled with gravel to reduce vibrations, and carefully mounted in industrial plants not accessible to everyone, for obvious safety reasons.

These CNC machines are essentially robots that can move tools in space to create objects. In subtractive technologies, the tool being moved may be a laser beam, a blade, a supersonic water jet, or a rotating instrument…the important thing is for it to be suitable to remove material. The digital revolution couldn't help taking CNC machines into consideration, so some years ago a series of open source projects appeared on the Net to help makers manufacture CNC machines in their own garages.

Obviously, these home-brewed machines aren't as precise as industrial models (such as the ShopBot 5-axis CNC shown in Figure 13-1), but to meet a maker's needs they are more than sufficient. Some manufacturing brands have created low-cost "desktop" models, taking up just about 4 square feet.

Figure 13-1 *An industrial-grade 5-axis CNC (ShopBot Tools)*

The machines we are going to cover in this chapter are CNC milling machines. They use a drill-type tool equipped with a cutting head to dig out the material. Milling machines can work on many kinds of material: wood, Plexiglas, plastic, cardboard, printed circuits, foams, and metals. Of course, milling machines that work on metals have a very different structure, weight, and feature set from home-based milling machines used for balsa wood.

The simplest machine moves in three directions, namely along the x-, y-, and z-axes: for this reason it is called a *3-axis machine*. More complex and expensive machines use four, five, or more axes; some even have the ability to rotate around the main axes. The rotation axis around the x-axis is called the A axis; then there are the B and C axes for the rotations around the y- and z-axes. Some machines use robotic arms that are able to make very complex movements and reach any part of the piece that the machine is carving. Fortunately, it will not be you, but the software, that manages the whole process.

The space within which the working tool can be moved defines the working area of the machine. For most purposes, unless you want to manufacture customized furniture pieces or a prefabricated building, a small working area is enough. For example, the Mebotics Microfactory (*http://www.mebotics.com/microfactory.html*), shown in Figure 13-2, has a working area of 12×12×6 inches.

Even if the working area may appear limited, consider that, just as in 3D printing, the biggest projects can usually be split into smaller parts. You can and should think of ways to make the project modular, which will enable you to create the object by joining or wedging in the different components. If you search online, you can find many examples of how to create wonderful snap-fit joints. For our own "garage projects," a small working area is more than sufficient.

Figure 13-2 *The Mebotics Microfactory at work*

The milling tool (the *endmill*) is so similar to a drill head that an inexperienced maker might not be able to tell the difference. However, a drill head is used to make *holes*, and works vertically, with its tip making the cuts. In contrast, most milling machines cut mainly on the edges, move on a horizontal plane, and remove material layer after layer.

Milling machines have molded heads with different shapes that are chosen according to the working needs (see Figure 13-3):

- Pointed-tip heads are used to carve or engrave.
- Ball-nose endmills are suitable to create rounded or "organic" surfaces.
- Square heads with one or more grooves ("flutes") are a compromise solution, able to work on both flat and curved surfaces.

What's the Use?

CNC milling machines carry out repetitive operations quickly and with great precision. They are used in all cases in which subtractive technology is more suitable than additive (for example, engravings, bas-reliefs, molds, cuts, and holes) and to work with materials other than plastic, like wood and metals, which you have to use when you need objects with high mechanical resistance.

Let's look at some examples of what you can create:

- Customized guitars, by engraving logos or names on existing guitars or by building all necessary parts from scratch
- Artistic joints—for example, the 3D version of the works made by M. C. Escher, engraver and graphic designer
- Molds for chocolate bars, as seen at the Fab Lab of Turin
- Wooden boxes and containers of different shapes and dimensions, engraved and decorated
- Molds for model building (e.g., airplane models)
- Different gearwheels and mechanisms, in metal
- Printed circuits: instead of photoengraving and forming them with acids, a milling machine head can engrave a plastic or copper sandwich
- Artistic objects

In general, a milling head is composed of a tapered stem with fixed dimensions, between 6 and 10 mm. The stem is tightened in the spindle, a retention system similar to that of conventional drills. The best spindles allow the head to be precisely inserted and always aligned: this is extremely important because milling cutters can work on fractions of millimeters, but if they are not perfectly assembled, the actual precision can be distorted and the tool's life expectancy may drop by 50%.

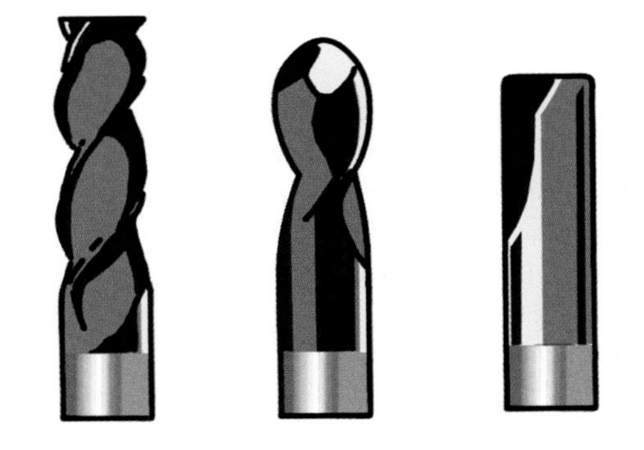

Figure 13-3 *From left to right, representation of a multi-flute, ball-nose, and single-flute endmill*

Many industrial machines are able to automatically change the tool on the spindle without an operator's intervention, thus guaranteeing the best precision in the replacement. The different working tools are kept in an automatized rack.

The CNC's control unit supervises the movement of the motors through circuits called *drivers*. To obtain higher precision, movements are measured with specific sensors called *encoders*.

Designing with a CNC

Knowing what a CNC can make will help you understand better what kind of drawing or design you are able to create.

A 3-axis milling machine can carry out three types of carving:

- 2D (Figure 13-4): The tool digs precise shapes with straight borders at a fixed depth, effectively working only on the plane of the x- and y-axes.
- 2.5D (Figure 13-5): The machine works again on areas parallel to the *xy* plane, but at different depths. The milling machine can cut profiles or dig. The surfaces created are only horizontal or vertical.
- 3D: The machine works simultaneously on three axes, so surfaces can have any kind of orientation.

Figure 13-4 *A ShopBot Handibot resting after some 2D carving*

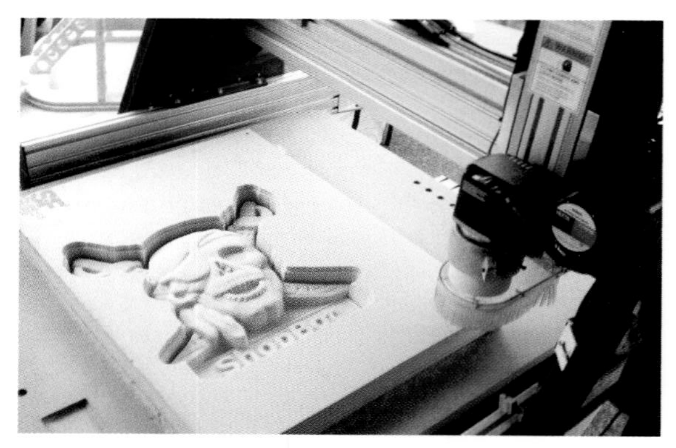

Figure 13-5 *2.5D carving on a ShopBot*

In the object designing and drawing step, it is important to bear in mind the features of the machine that will do the manufacturing.

The finished object resolution depends on both the features of the material and the dimensions of the tool being used. A 6 mm head may have some problems reproducing very tiny features; moreover, there may be the risk that, within the cut, some details get lost. The same problem occurs when the head doesn't have enough space to move (Figure 13-6), because moving there could ruin the part you're cutting.

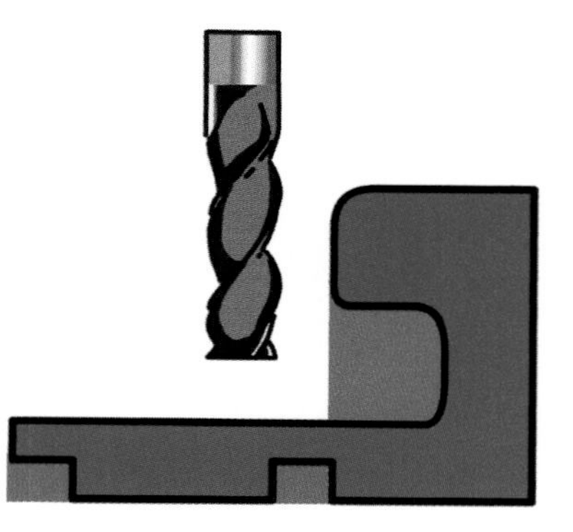

Figure 13-6 *The tool-path planning software (CAM) excludes all parts in red because they are not workable on a 3-axis machine*

A 3-axis machine is not able to rotate the material bring worked, and so cannot carry out operations on the side that's resting against the working surface. For this reason, control software eliminates all impossible work steps, thus avoiding damage to the workpiece.

Should you want to operate even on areas that cannot be reached in one step, you have to create systems of reference that allow you to turn the piece manually and realign the machine so that the process can continue. For this purpose, you can use holes and joints provided by pivots and alignment pegs, creating additional ones outside the "real" workpiece if necessary.

In addition, you have to consider the fact that the milling machine has a certain length (Figure 13-7) and that the spindle covers a quite large amount of space, so you can't dig wherever you want—for example, close to borders and walls.

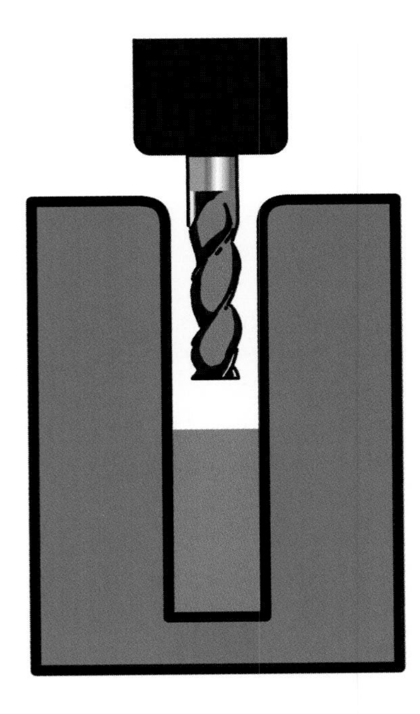

Figure 13-7 *The gap is too deep and you won't be able to dig it*

Software

Just as with 3D printing, milling requires three types of software:

- Computer Aided Design (CAD) software to create a model you will export in STL, EPS, or DXF format
- Computer Aided Manufacturing (CAM) software to make G-codes
- Software to control the machine

CAD Software

As with 3D printing, there is no particular software we recommend, so you should choose the one you like the best. You could look back at some of the 3D CAD packages from Chapter 11, but these CAD/drawing packages are worth checking out: Inkscape (*https://inkscape.org/*) is free and open source, and works well for 2D CAD; and Autodesk Fusion 360 (*http://fusion360.autodesk.com/features*) is free for some users (students, educators, startups) and you can use it for everything from 2D to 3D.

CAM Software

To turn a two- or three-dimensional model into a series of instructions for a CNC machine, you need CAM software that generates the toolpaths the machine must follow to create your object. The best software packages improve paths and have elaborate strategies to reduce working times and achieve better results. CAM software for milling machines is mostly commercial; it's extremely expensive and uses complex and old user interfaces. Thanks to the spread of designs for the self-manufacturing of milling machines, however, a series of software programs have appeared that are more suitable for a maker in terms of costs and complexity. We recommend FreeMill (*http://mecsoft.com/freemill*) (Figure 13-8) by MechSoft. FreeMill is a high-quality CAM for CNC; it is available for free, it uses quite simple algorithms, and it doesn't have complex optimization strategies. Still, it is a perfectly functioning software option derived from VisualMill, a more full-featured product.

FreeMill

Here's an overview of how you'd work with FreeMill.

Once the drawing is uploaded, you need to set the working parameters. On the left side of the window there is a *wizard* organized in cards on which you can set the main information. In the first step, include the axes orientation of your machine.

In the second step, indicate the dimensions and the margins of the block you intend to work on (Figure 13-9).

The CNC works on three axes and needs a point of reference or an axes origin. Here each machine is different; you can provide machine-specific information in this step (Figure 13-10).

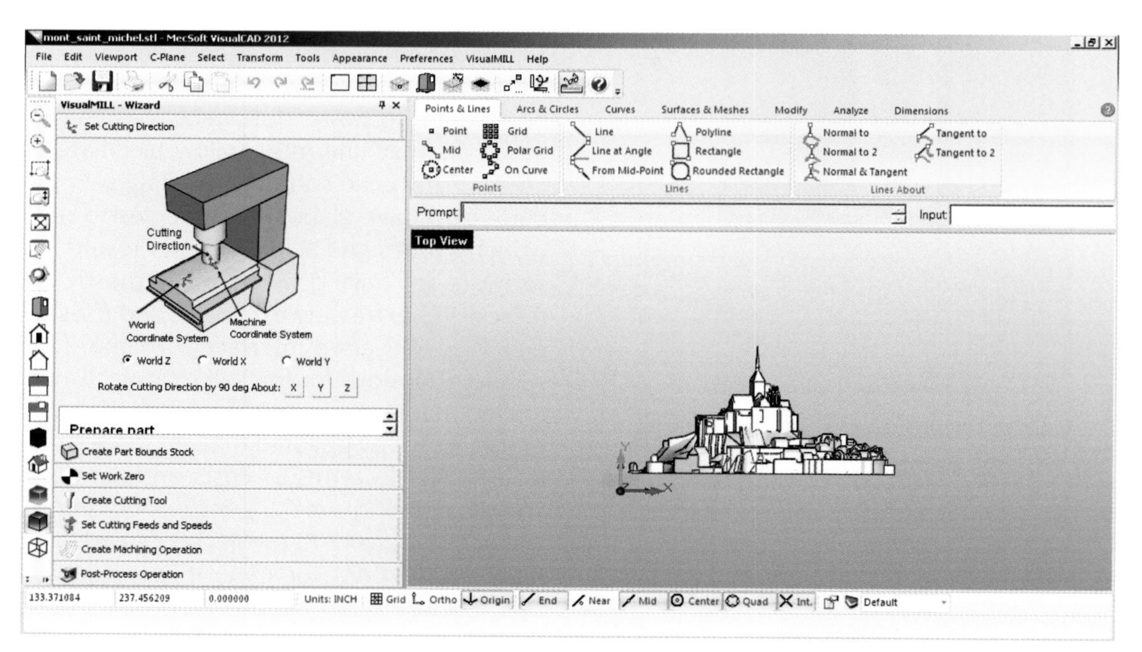

Figure 13-8 *The CAM FreeMill software by MechSoft*

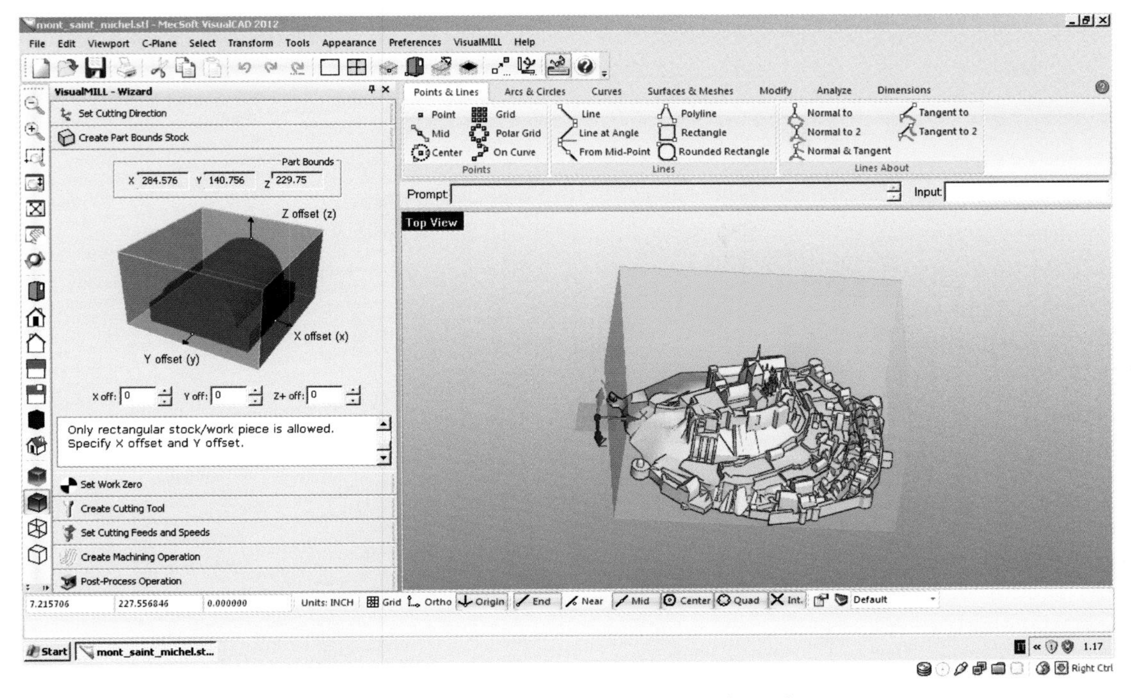

Figure 13-9 *Setting the working area and margins*

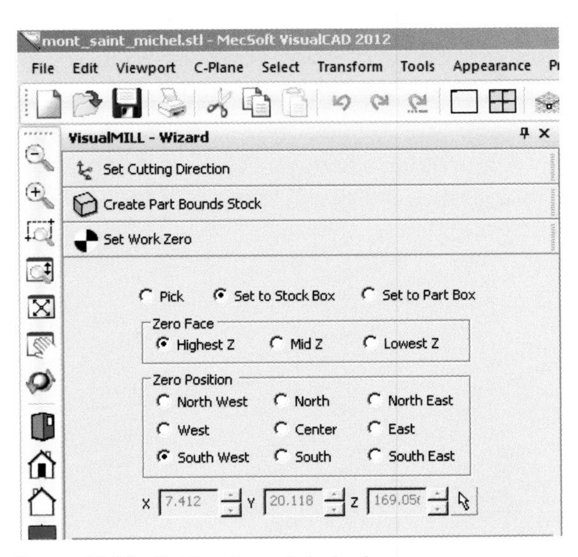

Figure 13-10 *Setting the point of reference*

In the next step, you will provide the type of tool (flat, rounded, or blunted head) and the corresponding dimensions, as well as the length and features of the spindle (Figure 13-11).

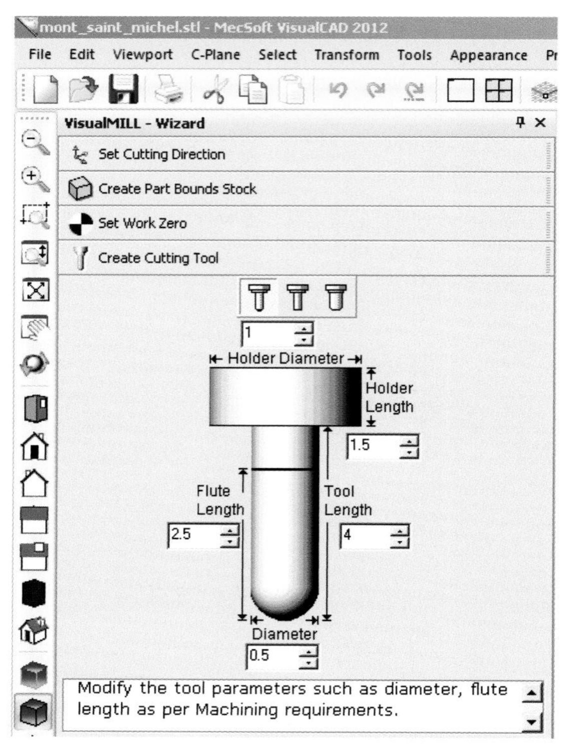

Figure 13-11 *Defining the tool dimensions*

Then, you have to specify the movement (feed) speed of the machine and the rotation speed of the tool (Figure 13-12).

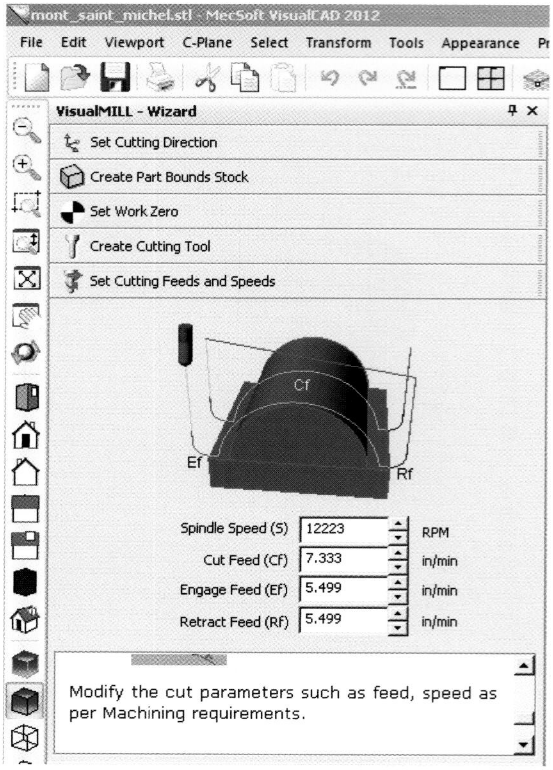

Figure 13-12 *Setting the speeds of the tool movement and rotation*

Now you're ready to set the final parameters to generate the G-code and visualize the work path. The panel allows you to indicate on which axis the main shifts will take place (x- or y-axis) (Figure 13-13).

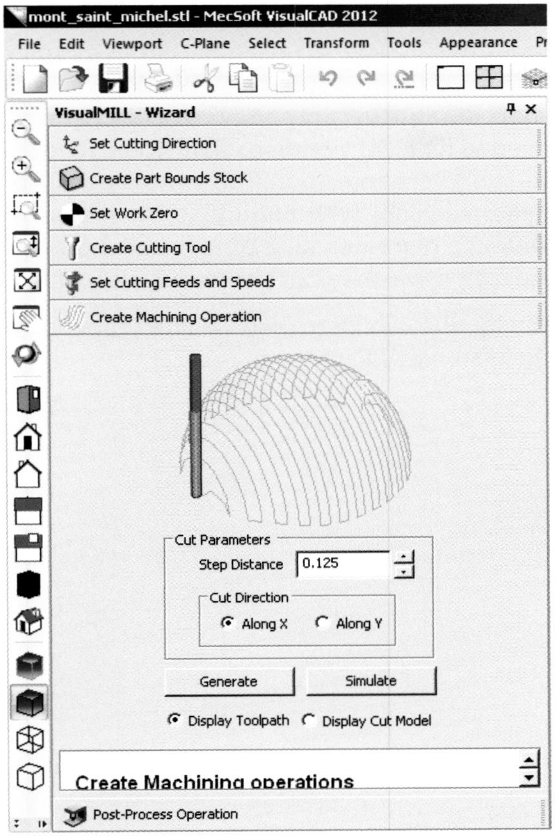

Figure 13-13 *Configuring the work paths*

The last step, the *Post-Process Operation*, is used to generate the file: you will find a long list of names among which you have to choose your control software. If you can't find it, you can use the *General Postprocessor* option.

Control Software

When you buy a professional milling machine, you usually receive matching software to operate it, called control software. If you build the machine by yourself, you will have to find or write your control software; we recommend a couple of options:

- MACH3 for Windows (*http://www.machsupport.com/software*) (Figure 13-14), created by ArtSoft USA; the free-of-charge version of the software has a 500-line G-code limit, while the complete version costs less than $200.
- EMC2 LinuxCNC (*http://www.linuxcnc.org*), an open source solution.

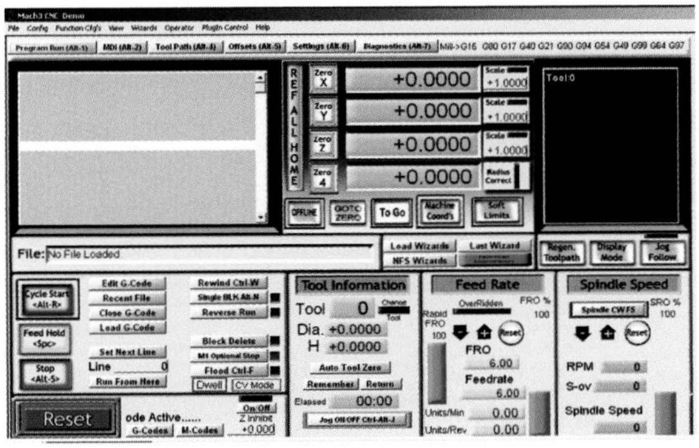

Figure 13-14 *The MACH3 control software*

Regardless of the software you choose, after installing it, you will have to configure it and set many parameters that vary for each machine—for example, the dimensions of the work area as well as the working and shifting speeds. You will then have to verify the signals received and transmitted and run some tests to fine-tune the system functioning and performance. Such an operation can take many hours of work and a lot of patience.

Does it sound complicated? It actually is!

Where Do We Turn?

Milling machines have not generated the same level of buzz as 3D printers, partly because they are more complicated. Moreover, the machines take up much more room, and they are harder to manage. However, milling machines offer a great deal of possibilities and are indispensable for some types of work.

If you are mainly interested in working on small objects, and in milling electronic circuits, it may be convenient to buy a small desktop milling machine (there are machines with build areas smaller than 10×10 cm).

If you need to work on bigger objects, you can turn to a Fab Lab, makerspace, or hackerspace, which are usually well equipped for this kind of activity. There, you can also find someone who will give you a hand with your creation or even make the object for you.

If you want to build a milling machine all by yourself (we have done it!), you can find a lot of information on the Internet; a good starting point is the site Build Your CNC (*http://www.buildyourcnc.com*), which also sells ready-made kits.

For objects that need very precise dimensions, or for metals, you will have to turn to a professional prototyping studio or a mechanic's workshop, because very big and heavy machines are needed to avoid vibrations that would jeopardize the level of precision. The same is true if you need machines with more than three axes. As all these machines take up a lot of room and are expensive, they can't be found everywhere.

Finally, the working costs mainly depend on the materials you choose.

Laser Cutting

14

The last digital manufacturing technique we are going to consider is *laser cutting*. Nowadays, it seems like without a laser cutter, you can't be a maker. Though we don't necessarily agree with that assessment, we do agree that a laser cutter is a very important tool in a maker's toolbox.

Lasers

The word *laser* is an acronym for *light amplification by stimulated emission of radiation*: big words that basically mean "watch out, this burns!"

A laser can direct a lot of power to a very small area, thereby melting, burning, or even vaporizing the material we are working on (or enemy robots). Based on this principle, laser cutters have been created that are capable of working on a wide range of materials.

Laser Cutters

There are various kinds of laser cutting machines on the market, which differ in the variety of materials they can work with, their mechanics, their optics, the kind of laser they use, and other factors. The most commonly available laser cutters are based on the excitation of a gas, mainly carbon dioxide (CO_2). Besides the optic system, which directs and focuses the infrared light emitted by the laser, an air or gas blast is used to clean the material from the debris generated by the laser cutting process.

On some cutters, the laser is fixed and the material moves, while on others the material stays in place and the laser beam moves. In most designs of this type, the laser *tube* remains stationary, and a series of movable mirrors convey the laser *beam* wherever needed. The mirrors move similarly to inkjet printer heads, though on two axes instead of one, so they can cover an entire plane. When you get right down to it, a laser cutter is essentially a computer numerical control (CNC) machine, which reads instructions and guides the tool in its operation.

A laser cutter can work with many materials: acrylic, paper and cardboard, wood, felt, fabric, rubber, leather, Plexiglas, and—if the machine is powerful enough—even some kinds of metal. The processing is also different: by adjusting power and focus you can cut the material, score it (for later cutting), or engrave it (by writing or drawing on it). See Figures 14-1 through 14-3 for three examples of laser cutting.

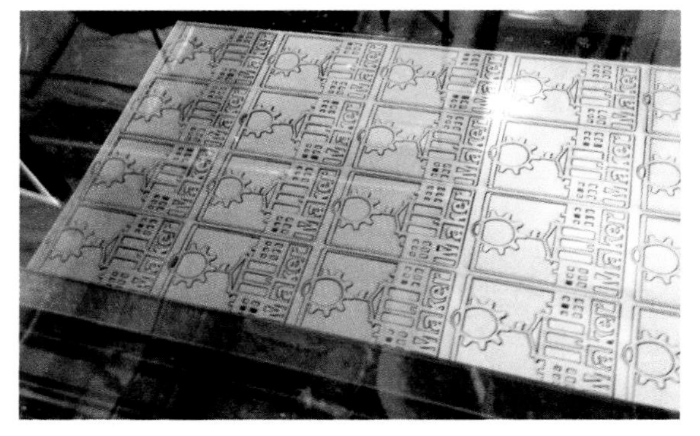

Figure 14-1 *Laser cutting attendee badges to commemorate the first White House Maker Faire*

Figure 14-2 *Finely detailed laser-cut work by Larry Zagorsky of AS220 Labs in Providence, Rhode Island*

The standard thickness of the materials is a few millimeters, though there are machines that can cope with dozens of millimeters. This shouldn't limit you to creating flat objects: you can design objects with cuts and joints that, when pieced together, create three dimensional objects, from simple boxes to more complex products.

Some software, such as Autodesk 123D Make, starts from a 3D model to generate flat shapes that, properly put together, can create actual sculptures.

Besides the many machines available on the market, there are community resources available, such as Lasersaur (*http://www.lasersaur.com*), an open source laser cutter. A machine of this kind is safe and powerful enough to meet the maker's needs, at a reasonable cost. Moreover, it is documented in all its aspects, so if you wanted to change something you would simply *fork* the initial project, clone the project documents repository, and use that as a starting point for your creation.

Even with open source alternatives, only a few of us can afford a laser cutter, given its relatively high cost. Fortunately, there are many services that put the full power of a laser cutter at your disposal in a click, and at a reasonable price. As of this writing (early 2015), they are so ubiquitous that a Google search for "laser cutting service" should turn up quite a few companies in your area. Many of them offer high-quality professional service and can help you in the modeling phase.

A laser cutter is one of the standard tools of a Fab Lab and in many makerspaces and hackerspaces, so if you are lucky enough to live in a city where there is one, you can join that space or sometimes pay a small fee to learn how to use the laser and access it for your projects.

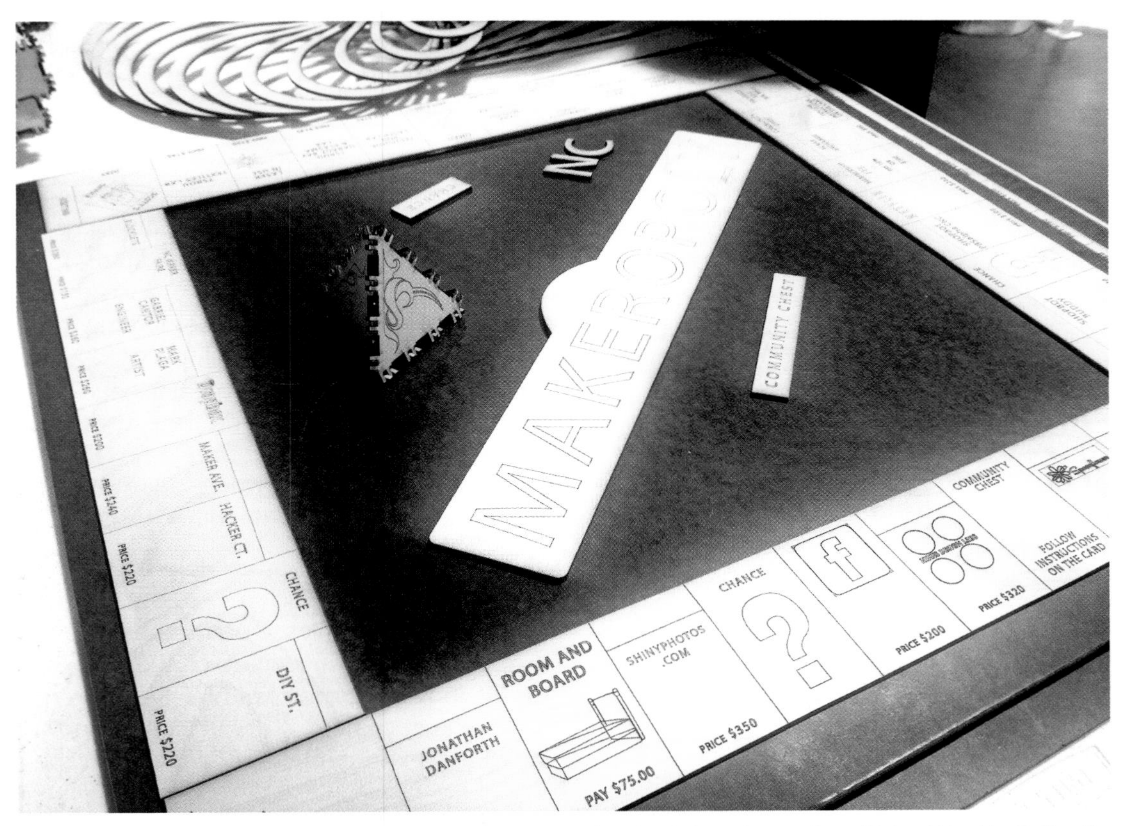

Figure 14-3 *Makeropoly, Mark Plaga's laser-cut homage to the classic board game from Maker Faire North Carolina*

What's the Use?

There are many things you can create with laser cutting:

- A perfectly functioning steadicam, made out of 3D printed parts, ball bearings, gym counterweights, and bicycle handles
- Snap-together boxes, customized and engraved
- Full-height plywood shapes resembling your friends
- Gears and cogs that are not subject to much wear or stress
- 3D objects, obtained by snapping together perpendicular sections or by layering parallel sheets
- Sculpted books
- Business cards
- Necklaces, earrings, pendants, buttons, and bangles
- Inscriptions or graphics engraved on different materials
- Artistic objects

Models

The model needed for laser cutting is basically a *vector* graphic in two dimensions, where the color of its lines refers to specific processing. The laser cutting software will translate this model into instruction that the machine can understand for the actual cutting process.

Vector graphics

When you look at pictures taken with a smartphone, you are looking at a *bitmap* or *raster* image, which is made of millions of tiny dots called *pixels*, each set to a different color.

When you use vector graphic software, you represent very complex shapes through mathematical relationships, just as you can draw a circle by knowing its center and radius. In a vector image, the software *calculates* the position of each pixel, based on the different formulas in the image file. Vector graphics can be more complicated, but they're also precise.

To prepare laser cut models, you can use any vector graphic tool—for example, *Adobe Illustrator* or *CorelDRAW*. You could also use *Inkscape*, a much less expensive tool (it is free!), which we'll describe next.

Inkscape

Download Inkscape from the official website (*http://inkscape.org*), follow the instructions for installing it on your operating system, and then run it. Inkscape is user friendly; among other things, it features a series of tutorials to guide you step by step in making something, even if you have never seen—let alone used—vector graphic software. Tutorials are designed to help you learn by doing, in perfect maker spirit: there are no complex and boring texts, just Inkscape files, which you can modify directly to understand how to use the different tools.

To open the tutorials, click Help on the menu, then highlight Tutorials. The system will display a list of available files. Open the first one, Inkscape: Basic. The first part of the tutorial explains how to move within the document, zoom the document in or out, create a new file, or edit an existing file. The default Inkscape file

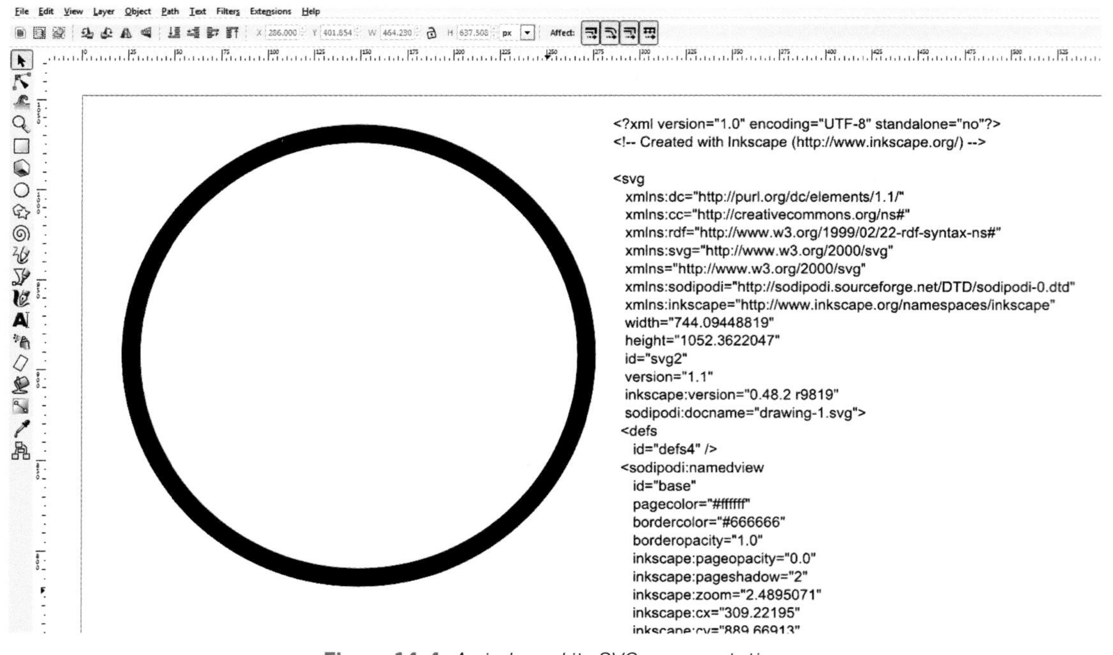

Figure 14-4 *A circle and its SVG representation*

format is *SVG* (Figure 14-4), which stands for *Scalable Vector Graphics*, a widely supported, standardized, and open format. SVG is based on XML, a text format, so Git (see Chapter 10) can help you keep track of the changes in your designs over time.

The circle definition looks like this:

```
<path
   sodipodi:type="arc"
   style="fill:#fe5b58;fill-opacity:
1;stroke:none"
   id="path2985"
   sodipodi:cx="151.42857"
   sodipodi:cy="295.21933"
   sodipodi:rx="271.42856"
   sodipodi:ry="277.14285" >
```

The tutorial guides you gradually while you create rectangles, stars, spirals, and other shapes. After playing with the tutorial for a while (we recommend the first two, at least), you can move on to the design of a mini-project: a simple sign with a couple of logos and some text. Create a new document and save it to a new folder/directory, which we'll refer to as the *working directory* of the project. Make sure to save as an uncompressed SVG, so that Git can show you differences later on.

Our logo (Figure 14-5) is made out of two images and two separate texts. Let's start with the text: click the Text button or press F8 in Inkscape. Click anywhere in the document—you'll have time to find the right positioning for the writing later—and start typing, as shown in Figure 14-6.

Figure 14-5 *The logo we want to transfer from the bits to the atoms world*

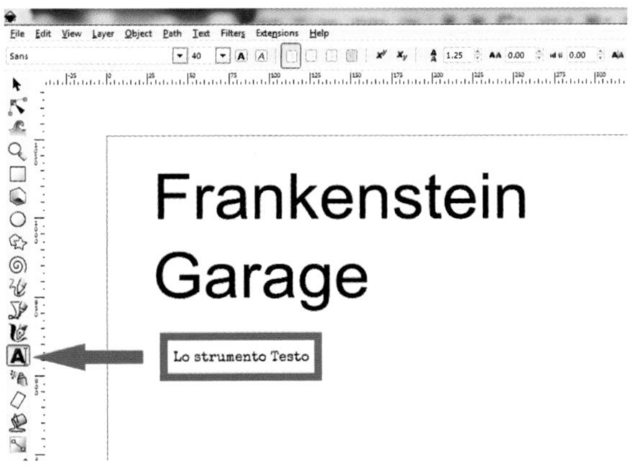

Figure 14-6 *I am not a Frankenstein. I am a Fronkensteen!*

To change the font, select the text and click the drop-down menu at the top left, then choose a font as shown in Figure 14-7.

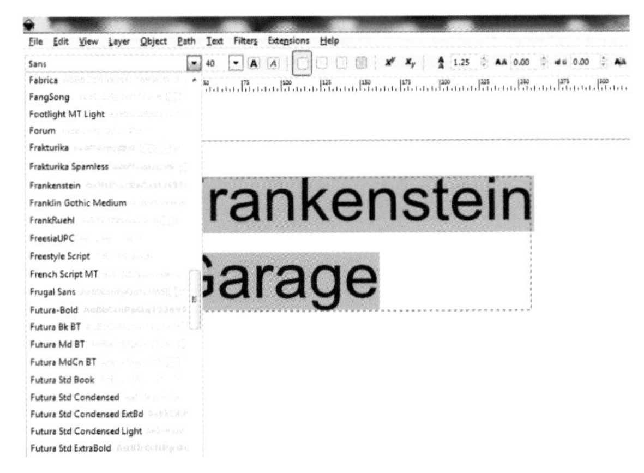

Figure 14-7 *Selecting a font for the text*

Using the other buttons of the toolbar available to you through the Text tool (Figure 14-8), you can further edit the text style, for example by changing character size and spacing.

Figure 14-8 *The edited text*

You can center the writing in the document: select the text, use the shortcut Shift-Control-A to open the Align and Distribute panel, select "Relative to Page," and click the button that has the tooltip "Center on vertical axis," as shown in Figure 14-9.

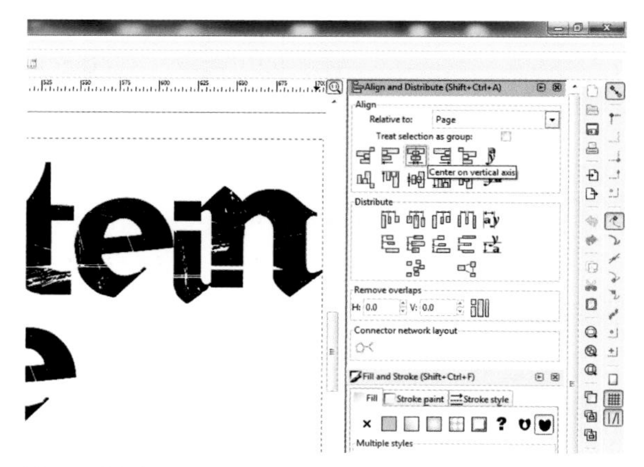

Figure 14-9 *Aligning objects*

Before using this text for a laser cut, you need to convert it to a vector path so that the outcome will be accurate on any machine. Be careful, because you can't undo this operation and you won't be able to easily edit the text any further. To be sure you have a backup, duplicate the layer that has the text and work on the copy. If you make a mistake, you can always throw it away and make a new copy of the original text layer. When you're finished, hide the original layer.

With the Shift-Control-L shortcut, open the Layers panel, double-click the layer name (Layer 1), and rename it something more descriptive, as shown in Figure 14-10.

Figure 14-10 *Renaming a layer*

Then, right-click the layer name and select Duplicate Current Layer. Rename this new layer, too. Renaming layers is not mandatory, but it is strongly recommended because it helps you figure out which object you're working on.

Hide the original text layer by clicking on its eyeball icon. Select the text in the duplicated layer and, from the Path menu tab, select "Object to Path." You can double-check the conversion by pressing F2, which allows direct editing of the various paths. See Figure 14-11.

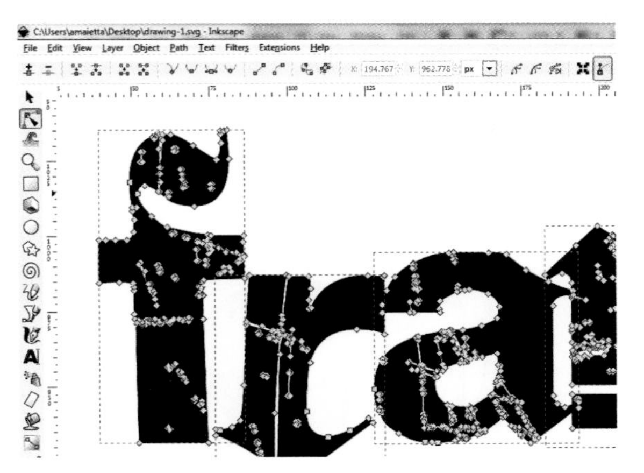

Figure 14-11 *The text, ready for cutting*

Laser printers translate vector line color to laser intensity. A fully black line, for example, indicates a fully powered laser cut. If you merely want the laser to add some shading to your material instead of cutting all the way through, you have to lighten the color of the line. Make sure all the text is still selected, then use the shortcut Shift-Control-F to open the Fill and Stroke panel. From the Fill tab, select a Flat Color RGB with values [128, 128, 128], which represents a medium grey, as shown in Figure 14-12.

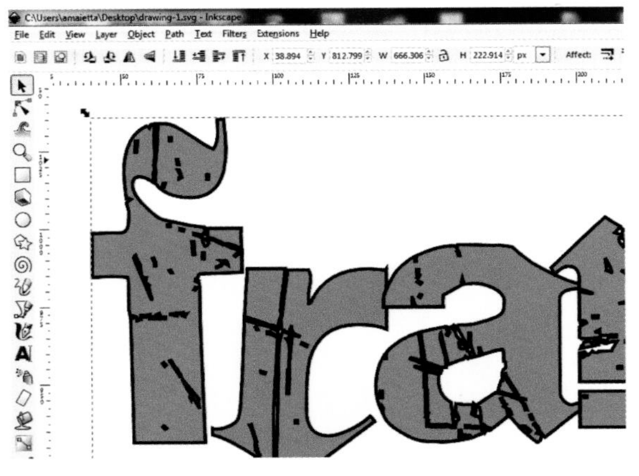

Figure 14-12 *Changing the text filling*

You have now modified the inner part of the text, but if you don't want each letter to be cut out, you still have to modify the outlines. Go to the Stroke Paint tab, select Flat Color, and set the strokes to RGB [128, 128, 128]. Then go to the Stroke Style tab and select 0.010 mm for the Width. You can see the result in Figure 14-13.

Figure 14-13 *The text, ready to be processed*

Now draw the outline of the sign: click the Rectangle tool, or press F4. (Inkscape has shortcuts for nearly everything, and the more you learn, the more productive you'll be.) Change the settings so that you have no filling and a full black outline, then draw a rectangle around the text. You can do this on a different layer if you like, for the sake of simplicity and clarity.

You can modify the corners of the rectangle by dragging the perimeter indicators as you learned from the tutorial (you did go through it, didn't you?) or editing the values from the toolbar boxes as shown in Figure 14-14.

Figure 14-14 *The sign outline*

Now let's move on to the images.

You can start with the one on the left (Figure 14-15), which you can find online (*http://bit.ly/1NyYQDo*). Download the image to your computer. From the File menu, select Import and insert the image into the Inkscape document. A dialog window will pop up, asking whether you want to embed the image in the document, or just link it. Choose embed, the default setting, and confirm. Scale and drag the image to its final position.

Figure 14-15 *The sign now has a logo too*

Once the image has been positioned, you have to turn it into a path, (i.e., a vector graphic), as you did with the text. Let's learn a new trick: image tracing. Select the logo and, from the Path menu, select Trace Bitmap. Different images need their own settings; in this case, you are interested only in the outlines, so you can copy the configuration from the ones in Figure 14-16. Click OK, then close the Trace Bitmap window.

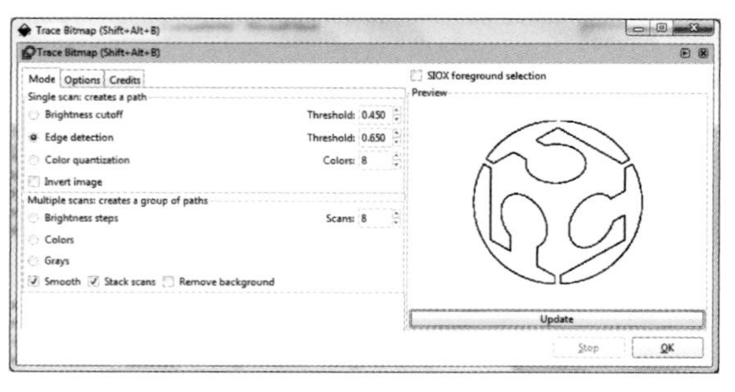

Figure 14-16 *Transforming the image into a path*

But wait! There are now *two* logos—one in full color, and one in black and white, on top of each other. You have to discard the colored logo, which you no longer need. Drag it away (Figure 14-17) to avoid any mistakes and press the Delete or Del button.

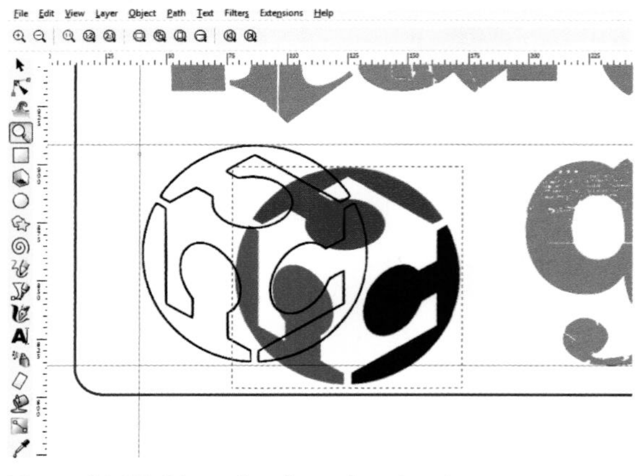

Figure 14-17 *Discarding the colored part*

If you have the two paths of the outline remaining, rather than just one, select the inner one and discard it.

You now need to follow two separate paths, because each side of the image will have a different filling. Click Path and select Break Apart (Figure 14-18).

Figure 14-18 *The separated paths of our logo*

Now, select each one of the three parts and choose Object→Fill; then, set the Fill for one shape to [101, 101, 101], the second to [53, 53,

53], and the third to [93, 93, 93]. Done! It wasn't hard, was it? Figure 14-19 shows the design so far.

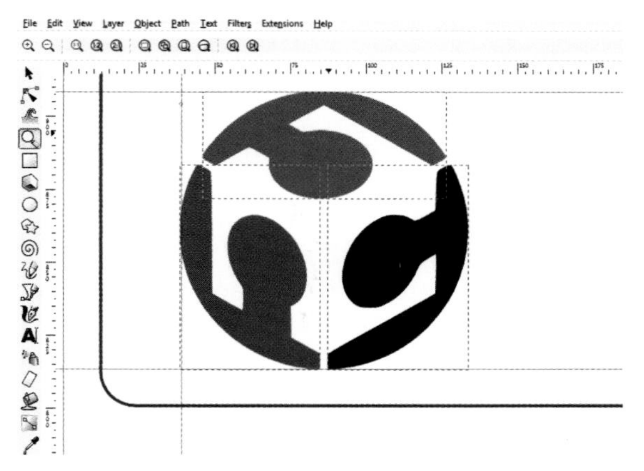

Figure 14-19 *The logo, ready for cutting?*

Now we only need the last image…

We could repeat the last procedure, but let's try to do this all manually, to help you get familiar with the tools. For the black you have in the original image, you'll use a deep raster incision, which could correspond to the black on the vector graphic. Create two rectangles with rounded angles; they will be Frankie's face. Set the fill to none and the stroke to a 4px (or so) solid line, as shown in Figure 14-20.

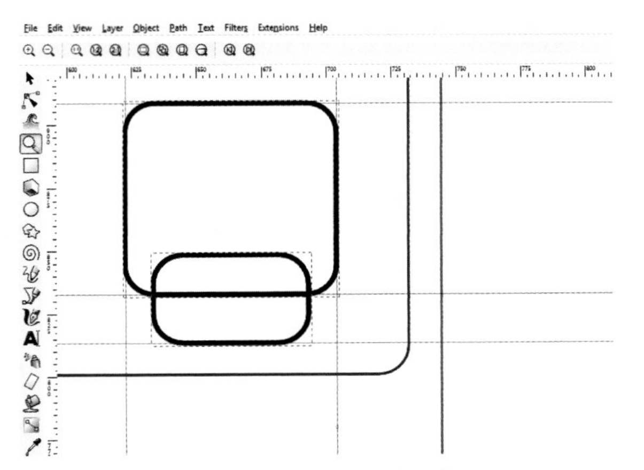

Figure 14-20 *The logo, ready for color filling*

Of course, the two rectangles are no good like this: we will keep only the outline. Select them both, and then from the Path menu, choose Union and obtain a single shape, as shown in Figure 14-21.

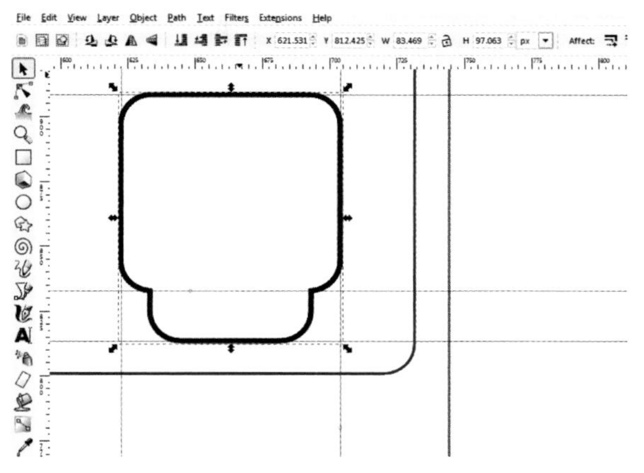

Figure 14-21 *The face outline*

For the hair you can use the Bézier tool for lines and curves, with the shortcut Shift-F6, and then draw the outline without worrying about what happens around the head, as shown in Figure 14-22.

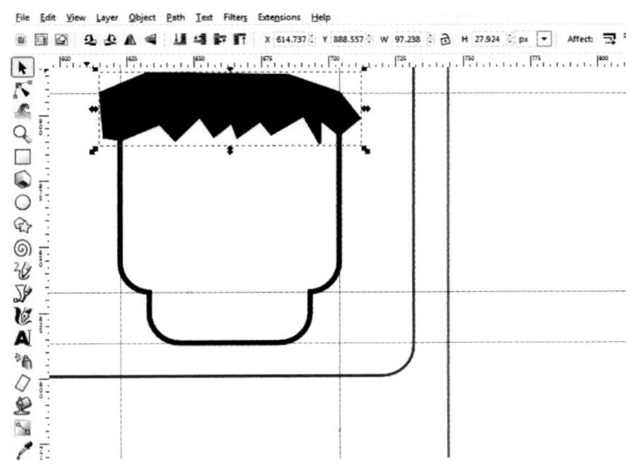

Figure 14-22 *Before the hairdresser*

Select the head and hair, and then from the Path menu, choose Division. Select the hair and change the filling (Figure 14-23), and now you're a barber!

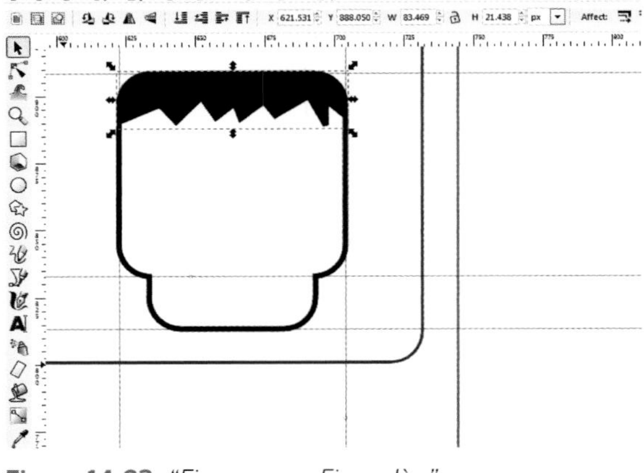

Figure 14-23 *"Figaro qua... Figaro là..."*

The eyes are very simple: first draw a circle with a black outline and white filling, then position another black circle within it. Copy both circles and resize the outer one. Now onto the nostrils, which are two simple white circles, and the scar, which is made of a horizontal line crossed by a diagonal one that's copied several times and evenly distributed. For these steps, use the Align and Distribute panel again.

For the mouth, draw a rectangle with rounded corners, and overlap it with another rectangle. Then, from the Path menu, select Difference, and then select the top corners of the mouth, click "Break path at selected nodes," and finally delete the horizontal line. Now your file is ready for cutting! (See Figure 14-24.)

Now that you are an Inkscape master, you can go crazy with your creations. But before you do, read the next section.

Optimizing a file

The cost of a laser-cut artifact depends on many factors, not least of which is the material itself, and the time needed for cutting. To reduce your costs, you can use *nesting* techniques to minimize unused space and optimize the cut. When you work at this kind of thing, you have to be careful and avoid double cutting lines, which originate when two shapes brought together end up with one line in common. This could have a negative effect on the material and also would be a waste of energy, since the cut would be carried out twice. To remove all double lines, you can use the same technique you used for Frankie's mouth in our example.

We all know that if you're using a saw to cut wooden boards, you need to adjust the measurements because the saw is going to "eat up," or "waste," a few millimeters: the same happens with a laser cut, though on a smaller scale. The material that gets lost with cutting is called *kerf*. In laser cutting, the amount of kerf depends, among other things, on the material, its

Figure 14-24 *The model, ready for cutting*

thickness, the kind of laser, and the lens and its focus. Roughly, we are talking about kerf of a few tenths of millimeter. Of course, the good old rule of "measure twice, cut once" is also valid for laser cutting. You have to take that into account when you design your project, especially if you have joints, which need to be as precise as possible.

The laser slows down along curves, so straight lines are faster. Also, it is worth placing the longest straight lines parallel to one of the laser cutter's axes, so that during cutting only one motor will move.

For laser cutting, as with almost any other maker's activity, an iterative approach is crucial: before cutting the material you want, make a cardboard or paper prototype, even on a smaller scale, to fix possible mistakes and spend a fraction of the cost to make the finished product. It sounds like a waste, but you won't regret it.

Tricks for 3D

As we said earlier, you can make 3D objects with a laser cutter by combining flat, laser-cut shapes. To manage joints in the best possible way, there are some tricks you can use. They help you consider kerf, unevenness of the material, and so on.

Moreover, each material has its own peculiarities; when you design joints for plastic objects, for instance, it is a good idea to include a small circle at the corners, called a *relief groove*, to better release forces and avoid any breakage of the material. Figure 14-25 shows a diagram from one of Ponoko's laser cutting tutorials (*http://support.ponoko.com/forums/345641-Laser-Cutting-Tutorials-Tips*). Ponoko offers laser cutting, among many other services.

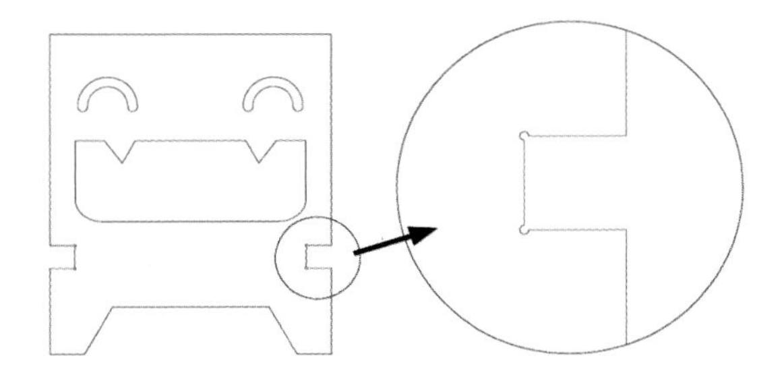

Figure 14-25 *Careful with those corners! (Ponoko)*

There are many other examples to be found on the Internet and even more sources of inspiration. Now you have all the components you need to create wonderful objects...you only need to learn how to make them come alive!

PART IV

Giving Life to Objects

Life, do you hear me?
Give my creation life!

—Dr. Frederick Frankenstein

- Chapter 15, Electronics and Fairy Dust
- Chapter 16, Arduino
- Chapter 17, Expanding Arduino
- Chapter 18, Raspberry Pi
- Chapter 19, Processing
- Chapter 20, The Internet of Things

Electronics and Fairy Dust

15

Tinker Bell and Peter Pan could fly thanks to a happy thought and some fairy dust. Our projects also need some fairy dust to fly, in the form of ghostly subatomic particles called *electrons*. Moving electrons are the basis of electricity, and their behavior in components is the basis of the field we call electronics. A beginner might view electronics as a difficult and inscrutable topic, and circuits as tiny and complicated objects. Electronics books are chock-full of math and physics formulas, so there is apparently nothing magic or fantastical...just a bunch of stuff an engineer would love!

But if you looked at electronics metaphorically, with an artist's eye, you could see wires and currents as pipes and water, transistors and potentiometers as valves and taps...and you would find out that electronics can be pretty intuitive, and a lot of fun too!

Hello World!

When a programmer approaches a new computer language, she first checks that everything is properly installed on the computer, and that the development environment can compile her program. For this first program, it's a programming tradition to write one called *Hello World*. Hello World does nothing more than print the words "Hello World" on the screen. Ridiculously simple, yes, but successfully writing, compiling, and running the program tells you that your development environment works.

With electronic circuits, there usually isn't a screen to display anything on. So hardware hackers have developed their own tradition: they add an LED to the circuit. If the LED lights up or blinks in a controlled manner, the circuit works.

What You Need

Let's take a brief look at the list of materials you need. To light up an LED, the very first thing you need is...an LED. LED stands for light emitting diode, and in true engineering fashion those words describe it completely: it is a diode (a component that lets electricity flow through it in only one direction) that emits light. As you'll see later when it comes to building the circuit, the diode function of LEDs is very important. LEDs are relatively cheap; they can be purchased in bulk online for as little as 2 cents each. They're available in nearly every color of the rainbow and beyond, from infrared to ultraviolet.

The second thing you need is a resistor. True to its name, a resistor *resists* the easy flow of electricity. Think of it as a sponge embedded in a hose, changing the flow of water (or electricity) from a torrent to a trickle. Resistors have different resistance values, from very weak to very strong, which are coded onto their surface in the form of stripes of many colors. For the purposes of this project, you'll need a resistor that is striped orange, white, and brown. The fourth band is less significant, but is likely to be gold or silver. These colors indicate a resistor valued at 390 ohms, but you can use a 1K ohm resistor (brown, black, and red).

If you have a five-band (blue) resistor, the first four bands of a 390 ohm resistor will be orange, white, black, and black. The first four bands of a 1K ohm five-band resistor will be brown, black, black, and brown.

To power the circuit you need some kind of energy source; the simplest solution is a 9 volt battery with a snap connector to wire it to the components. Finally, you will need some thin-gauge electric wire to connect the parts together, and some insulating tape to connect everything.

You can find everything you need for this project (except for the battery and tape) in Maker Shed's Mintronics: Survival Pack (http://www.makershed.com/products/mintronics-survival-pack). It includes the LED, the battery snap, and the resistor—among many other components you'll find useful.

A First Circuit

Start from the battery snap. If you buy your battery snap from an electronics retailer or online store like Maker Shed, the ends of the wire are probably already stripped. If you salvage the battery snap from a broken toy or radio, you'll need to strip the ends yourself. Don't use your teeth for stripping the wire! Use a dedicated wire stripper tool for this.

Twist the red wire from the battery clip around one of the resistor pins (either pin is OK; a resistor is not a diode, so power can run through it in any direction) and wrap up the connection with some insulating tape to make the joint more stable. Cut off a few inches of narrow-gauge electrical wire, and strip the insulation from a half-inch of both ends. Twist one end of the wire around the other resistor pin, and secure it with insulating tape. Right now, what you've got is simply an extension of the battery pack's red wire, with a resistor added to it. Spread the LED's pins, twist this wire around the longer LED pin, and secure it with tape. Finally, connect the shorter pin of the LED to the battery clip's black wire. Figure 15-1 shows how the circuit is wired up.

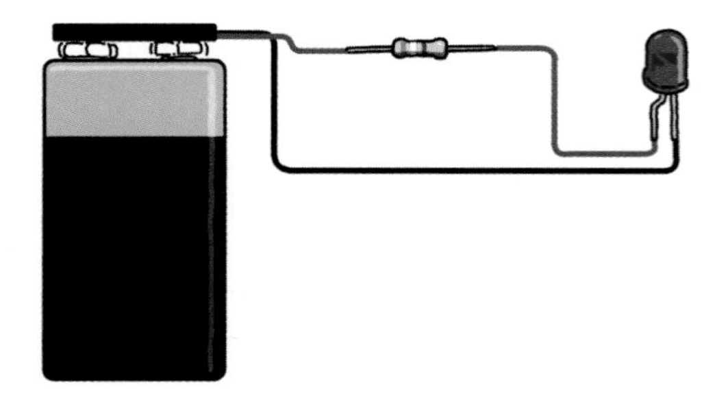

Figure 15-1 *Our very first circuit*

Now you just need to connect the battery and...Hello World!

> **Did You Burn Out Your LED?**
>
> If, instead of a happy glowing LED, you get a popping or snapping sound followed by a puff of smoke and an unpleasant smell, then you've just learned something important: you need to make sure the resistor's connected properly before applying power. If you omit the resistor, or accidentally bypass it with untidy wiring, the LED will receive too much current and burn out.
>
> If you are tempted to experience the destruction of an LED firsthand, take a few precautions: don't hold the LED in your hand when you try it, wear safety glasses, and perform this test in a well-ventilated area.

Current, Voltage, and Resistance

What do water and electricity have in common? You'd be surprised. Water can help us understand how electricity works, because we can explain many phenomena in a simple way by comparing the two. The metaphor is neither new nor original, and it is often used by those who teach electronics.

Let's first think about where electricity begins. Inside an atom (Figure 15-2), protons and neutrons make up the atomic nucleus, and electrons revolve around the nucleus. Electrical phenomena derives from the electrons' properties. Protons have a positive charge, electrons have a negative charge, and neutrons have no charge.

In metals, electrons are free to hop away from their nucleus, and move around within the metallic material. When there's a difference in electrical potential between one end of the metal and another, the electrons flow (at the speed of light) through the material. If you were to connect a battery directly to a piece of metal, the electrons would move at such a high flow rate that the metal (and the battery) would get hot very quickly, and continue to heat up until the battery is discharged, or worst case, until the battery or metal catches fire.

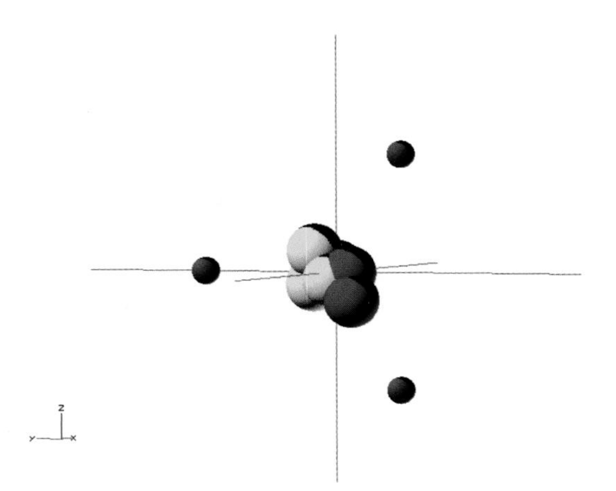

Figure 15-2 *A lithium atom drawn with OpenSCAD: the nucleus is composed of three protons (in red) and three neutrons (in yellow); the three electrons (in blue) revolve around the nucleus*

When we speak of electrical phenomena, we have to consider three main physical quantities: current (the flow rate just mentioned), resistance (something that can restrict that flow), and voltage (which we'll get into shortly).

Current

When you sit on a river bank, you see that water moves downstream from its source to the sea. We associate the term *current* with this movement. Similarly, when you attach a garden hose to a tap, water moves inside the hose and generates a (water) current.

The behavior of an electrical circuit is similar (but not identical; metaphors can only stretch so far). Instead of water molecules flowing through a hose, an electrical circuit has subatomic particles called *electrons* flowing through a wire. Wires are made of materials, usually metals, that allow the transfer of electrons from one atom to another. Materials of this kind are called *conductors*, while those that don't allow electrons to flow are called *insulators*.

The intensity of the electric current can be compared to the flow rate of a river. In the Nile, the quantity of water flowing in a specific time through a section of the river is much higher than what flows through, say, an irrigation canal. In the same way, the electric current flowing into an electric oven is much higher than the current in a smartphone.

The unit of measurement of electrical current is the *ampere* (A), or amp for short, from André-Marie Ampère, one of the main researchers of electromagnetism. The tool used to measure the strength of an electrical current is the *ammeter*, shown in Figure 15-3.

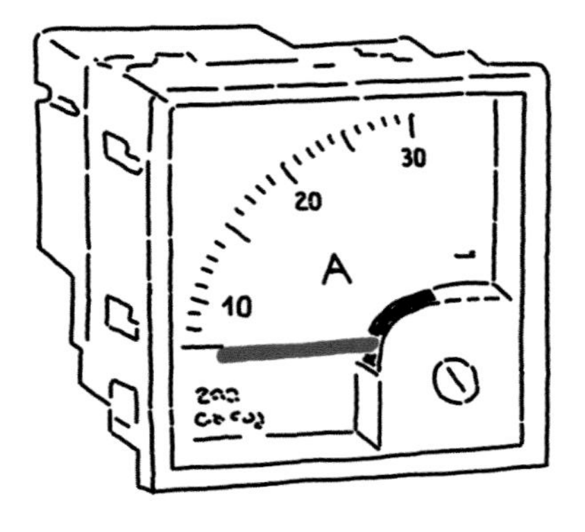

Figure 15-3 *An analog ammeter*

Table 15-1 shows the typical current consumed by some common devices.

Table 15-1 *Current used by some devices*

Device	Current draw in amps (A)
Train or tram	100–500
Oven	10–20
Clock radio	1
MP3 player	0.1

Voltage

Keeping the comparison with water, we see that in nature water flows only when there is an incline. We also notice that the intensity with which water falls depends on the height of the incline: consider the small inclines we find in streams and compare them with the Niagara Falls! Voltage can be compared to the incline from which water falls, or to the pressure of water in a pipe.

If we filled up a very long pipe with water and placed it horizontally on the floor, water would come out of the ends, though with little intensity. But if we lifted one end of the pipe, water would come out from the other end at a higher intensity, or, more correctly, with a stronger pressure, and this pressure would increase if the incline were steeper.

The unit of measurement of electrical "pressure," or voltage, is *volt* (V), from Alessandro Volta, inventor of the very first batteries and a pioneer in the study of electrical currents. The tool used to measure voltage is called a *voltmeter* (Figure 15-4). Table 15-2 lists the voltages required by some common electrical devices.

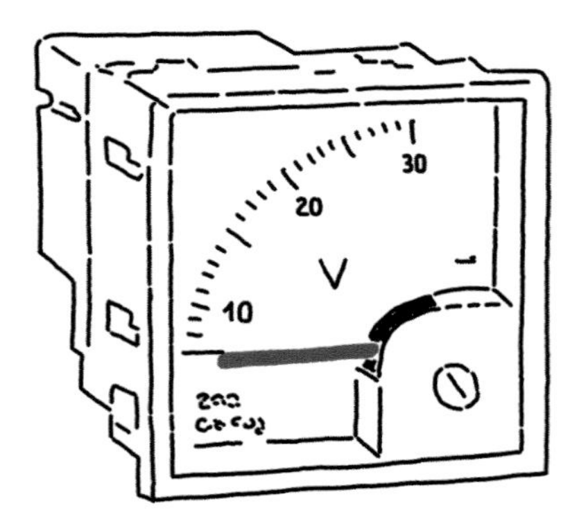

Figure 15-4 *An analog voltmeter; it doesn't look too different from the ammeter*

Take Care with Electricity

Electricity can kill. Most of the voltages and amperages you'll use in common maker projects are safe to use, but it doesn't take a lot of electricity to be dangerous. The effects caused by an electric current crossing your body can include:

- Muscular contractions or paralysis
- Breathing difficulties
- Suffocation (can cause death)
- Heart fibrillation (can cause death)

The Occupational Safety and Health Administration's publication, How Electrical Current Affects the Human Body (*http://1.usa.gov/1NzOQvt*), includes a chart of the dangers as well as the conditions (such as moisture, voltage, and exposure time) that can harm or kill.

Table 15-2 *Different electrical devices require different voltages*

Device	Required operating voltage (V)
Train or tram	3,000
Kitchen oven	220
Clock radio	12
MP3 player	3

Resistance

Imagine you are watering your flowers with a hose, when suddenly the water flow decreases (Figure 15-5). Someone may have stepped on the hose, thus reducing its cross-section, so it's harder for the water to flow through. Electrical resistances are composed of materials that make the transit of electrons difficult, similar to the foot on the hose. Resistance is therefore a quantity that measures the blockage of an electrical current. The more the material prevents electrons from passing, the higher the resistance value. Conductors have very low resistance, while insulators are characterized by very high resistance.

Figure 15-5 *When you step on a hose, you prevent water from flowing easily*

Resistance is measured in ohms (Ω), named for Georg Simon Ohm, who was the first to identify the directly proportional relationship between the resistance applied to a conductor and electrical current: this important relationship is called *Ohm's law*.

Just as the resistance in a hose increases if the foot stepping on it is bigger, electrical current increases if the material that electrons must flow through is *longer*. For each material we can indicate a specific electrical resistance as shown in Table 15-3, which defines how much 1 meter of material with a 1 mm^2 cross-section opposes the transit of electrons. The specific resistance is measured in ohm/meter (Ω/m). Metals are great conductors, which means they have very low resistance, while air and glass are excellent insulators.

Table 15-3 *Some examples of specific resistance*

Material	Specific resistance (Ω/m)
Copper	0.0000000169
Iron	0.0000000968
Human body	2,000
Glass	10,000,000,000

Circuits and Components

When we connect different electrical *components* in a closed path in which the current can continuously flow, we create a *circuit*.

Circuits

Although the underlying physics is somewhat complex and quite interesting, the flow of electricity in a circuit has enough in common with a hydraulic circuit (Figure 15-6) that we can model its behavior more simply.

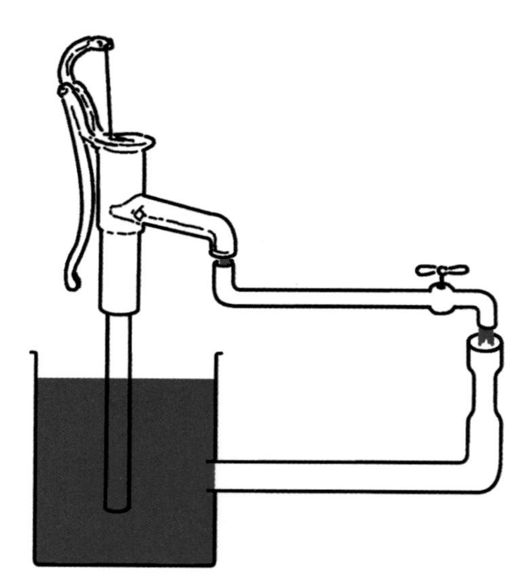

Figure 15-6 *A simple hydraulic circuit*

The most important rule of an electrical circuit is *Kirchhoff's current law*. Simply put, this states that the sum of the currents entering the circuit is equal to the sum of the currents going out of the circuit. (To use the analogy of a hydraulic circuit, this means that all water flowing into the circuit from the tap flows out at the other end.) That's why there are no electrical components with only one terminal: if the current flows into a component, it must also flow out of it; otherwise, the circuit won't work.

Whoa, What About an Antenna?

An antenna seems to break Kirchhoff's current law, since it has only one terminal. But an antenna is a case of its own: when an antenna receives a signal, current is induced in the body of the antenna, and it flows out through the antenna terminal, much the same way a rain bucket collects water that falls from the sky and can be drained off through a spigot.

Components

Each type of electronic component has a part number that is used worldwide: a type 2N2222 transistor is a type 2N2222 transistor anywhere you go. Where possible, the code is stamped on the component body using numbers and letters. Where it is not possible (or due to tradition), the code takes the form of notches, grooves, or colored stripes on the body of the component.

For most components, or family of components, manufacturers provide short manuals of instructions for free, called *datasheets* (see Figure 15-7), with all required information and warnings about how to use the components in the proper way. These datasheets can include electrical parameters, measures and dimensions, warnings, and sometimes even small demonstration circuits.

P2N2222A

Amplifier Transistors

NPN Silicon

Features

- Pb–Free Packages are Available*

MAXIMUM RATINGS (T_A = 25°C unless otherwise noted)

Characteristic	Symbol	Value	Unit
Collector – Emitter Voltage	V_{CEO}	40	Vdc
Collector – Base Voltage	V_{CBO}	75	Vdc
Emitter – Base Voltage	V_{EBO}	6.0	Vdc
Collector Current – Continuous	I_C	600	mAdc
Total Device Dissipation @ T_A = 25°C Derate above 25°C	P_D	625 5.0	mW mW/°C
Total Device Dissipation @ T_C = 25°C Derate above 25°C	P_D	1.5 12	W mW/°C
Operating and Storage Junction Temperature Range	T_J, T_{stg}	–55 to +150	°C

THERMAL CHARACTERISTICS

Characteristic	Symbol	Max	Unit
Thermal Resistance, Junction to Ambient	$R_{\theta JA}$	200	°C/W
Thermal Resistance, Junction to Case	$R_{\theta JC}$	83.3	°C/W

Stresses exceeding Maximum Ratings may damage the device. Maximum Ratings are stress ratings only. Functional operation above the Recommended Operating Conditions is not implied. Extended exposure to stresses above the Recommended Operating Conditions may affect device reliability.

ON Semiconductor®

http://onsemi.com

COLLECTOR 1

2 BASE

3 EMITTER

TO–92 (T0–226AA) CASE 29–11 STYLE 17

MARKING DIAGRAM

P2N2
222A
AYWW •

P2N2 = Device Code
222A = Specific Device
A = Assembly Location
Y = Year
WW = Work Week
• = Pb–Free Package

(Note: Microdot may be in either location)

ORDERING INFORMATION

Device	Package	Shipping†
P2N2222A	TO–92	5000 Units / Bulk
P2N2222AG	TO–92 (Pb–Free)	5000 Units / Bulk
P2N2222ARL1	TO–92	2000 / Tape & Ammo
P2N2222ARL1G	TO–92 (Pb–Free)	2000 / Tape & Ammo
P2N2222AZL1	TO–92	2000 / Tape & Reel
P2N2222AZL1G	TO–92 (Pb–Free)	2000 Units / Tube

†For information on tape and reel specifications, including part orientation and tape sizes, please refer to our Tape and Reel Packaging Specification Brochure, BRD8011/D.

*For additional information on our Pb–Free strategy and soldering details, please download the ON Semiconductor Soldering and Mounting Techniques Reference Manual, SOLDERRM/D.

© Semiconductor Components Industries, LLC, 2006
March, 2006 – Rev. 3

1

Publication Order Number:
P2N2222A/D

Figure 15-7 *One page of the Transistor 2N2222/D datasheet of Semiconductor Components Industries, LLC (used with permission from SCILLC dba ON Semiconductor)*

Where do you get components?

Up until about the mid-1980s, the United States was dotted with locally owned electrical and electronic repair shops. Throughout the history of electric appliances, when your toaster, radio, TV, hi-fi stereo, or any other electronic equipment broke down, it made economic sense to repair it (or to have it repaired) rather than throw it away. The local repair shop was where you could find new and used electrical components (sometimes painstakingly desoldered from irreparable machines), as well as someone who knew how to use them. When Radio Shack and other electronic parts retailers began their gigantic expansion at the end of the 1950s, they made electronic components even easier to obtain, and they also were careful to staff their stores with people who could answer your questions about diodes and transistors and such. The 1960s and early 1970s were a golden age of hobby electronics.

Things started to change in the early 1980s. It became cheaper to throw away a broken appliance and buy a new one, rather than have the old one fixed. Neighborhood repair shops started dying. Hobby electronics started dying. Fewer people knew how to solder. Electronic parts stores "rebranded" themselves as electronic gadget stores, and employees who understood resistors and capacitors were few and far between.

Fortunately, on the Internet you can find nearly anything you want. There are a lot of sites from which you can buy loose components. Web stores like Farnell (*http://www.farnell.com*), Element 14 (*http://www.element14.com*), Adafruit (*http://www.adafruit.com*), SparkFun (*https://www.sparkfun.com*), Jameco (*http://www.jameco.com*), Mouser (*http://www.mouser.com*), and DigiKey (*https://www.digikey.com*) sell and ship electronic components around the world. You can also find the datasheets for the components you buy right on these websites. If you want to purchase kits that contain a collection of parts, Maker Shed (*http://makershed.com*), Adafruit, and SparkFun have many choices. Sometimes these collections are an assortment of components that are good to have at the ready; sometimes they are kits for a specific project or purpose.

Of course, in real maker's style, you can cannibalize your old Christmas gadgets or your children's or grandchildren's broken and abandoned toys with just a little patience. You never know what you'll find, but with some luck you may come across something useful.

Now, let's have a look at some of the most important components, shown in Figures 15-8 and 15-9.

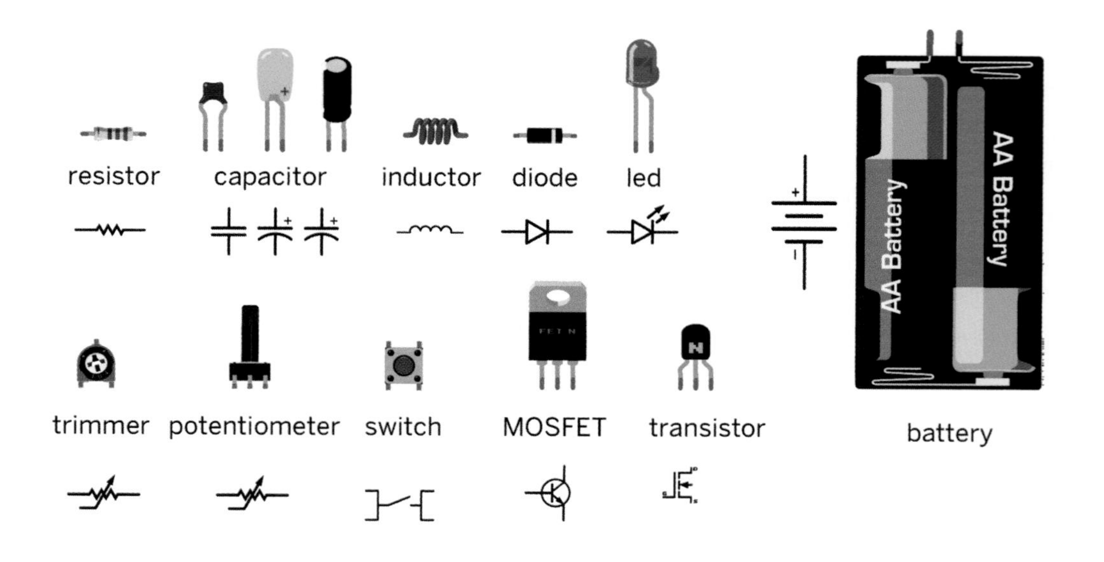

Figure 15-8 *An overview of the main electronic components with their corresponding symbols*

Many of the circuit diagrams in this chapter are created with Fritzing (http://fritzing.org/home/), an open source initiative to make it easy to design and manufacture electronic projects.

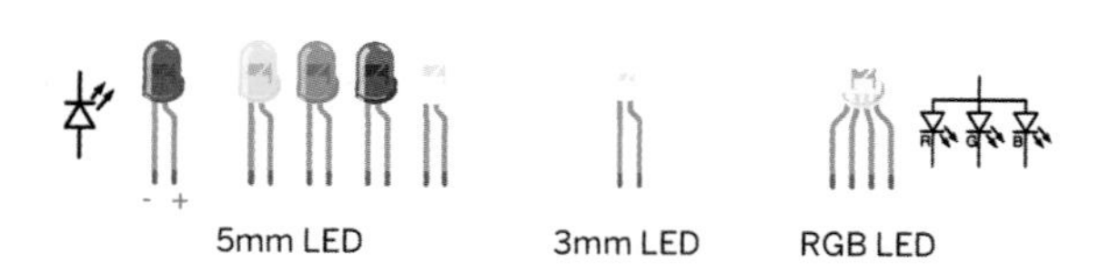

Figure 15-9 *Different types of LED with their corresponding electric symbols*

LED

As we explained earlier, an LED is a diode that emits light. It does this by exploiting a property called *electroluminescence*, where passing an electric current through certain materials causes them to emit light. Because electroluminescence doesn't generate heat the way an incandescent bulb does, LEDs generally don't get perceptibly warm unless they are extremely high-power, or unless you put too much current through them.

LEDs usually use a voltage between 1.6 and 3.6 V, and need a current ranging from 20 to 50 milliamps (mA), depending on the specific model of LED you're using. They are one of the few electronic components that don't have a code stamped on them, in part because their body is transparent or translucent.

Look very carefully at a standard 5 mm LED. Notice that one edge of the LED is blunted so that it is essentially flat instead of curved. That flat spot indicates the *cathode*, or negative terminal of the LED. The cathode is also distinguished by being shorter than the other terminal—you can remember this as the "minus" terminal because it has something "subtracted" from it.

There are also LEDs that can generate many colors; some multicolor LEDs have three terminals and can generate only red, green, or yellow light (by turning on green and red together). Four-terminal RGB LEDs are also available: three LEDs are contained in a single capsule with which it is possible to generate many colors. What more can we ask for?

Resistors

We can use some components, like the one we used in our first circuit, to resist the current and reduce its flow. These components are *resistors* (Figure 15-10), which are small, cylindrical objects, vaguely the size of a large grain of rice.

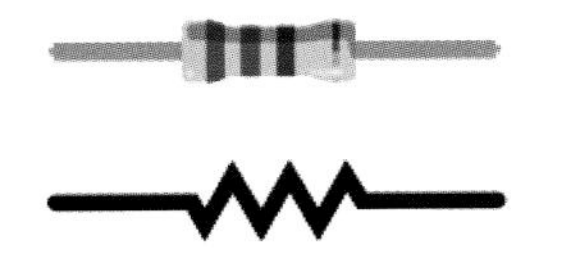

Figure 15-10 *A resistor*

Resistors have colored bands or rings printed on them. The specific colors, in specific order, specify the resistance value of the resistor. The code seems complicated at first, but it's not very difficult to pick up: at one end of the resistor there is a band indicating the tolerance (i.e., how precise the resistance value is): if it is gold, its value is 5%, and if silver it is 10%. When you read the resistance, keep this last band on the right: from left to right, each color corresponds to a number, as shown in Table 15-4.

Starting from the left, look at the first two bands and note the corresponding number; the third band indicates how many zeros to append to the first two numbers to obtain the actual value of the resistor. In our first example (Figure 15-1) we used a resistor with orange, white, and brown bands (the last band, the gold or silver one, does not factor into the value calculation). Orange corresponds to 3, white to 9, and brown to 1, so we'll write the number 39 followed by one 0: the resistance value is 390 ohms.

Table 15-4 *Resistor color codes*

Color	Numbers
Black	0
Brown	1
Red	2
Orange	3
Yellow	4
Green	5
Blue	6
Violet	7
Grey	8
White	9

What happens if you link two resistors, one after the other? Think back to the example of the hose with the foot stepping on it. If you step on the hose with both feet, you will not get just one obstruction, but two, resulting in a higher resistance. When you apply this principle to electronic components, it is called a connection *in series* and the resulting resistance is the sum of the single resistors:

$$R_{total} = R_1 + R_2$$

In Figure 15-11, you see two resistors, one with red (2), red (2), and brown (one 0), giving us 220 ohms; and the other with brown (1), black (0), and red (two 0s), giving us 1,000 ohms or 1 kilohm (abbreviated 1K). Together they total 1,220 ohms.

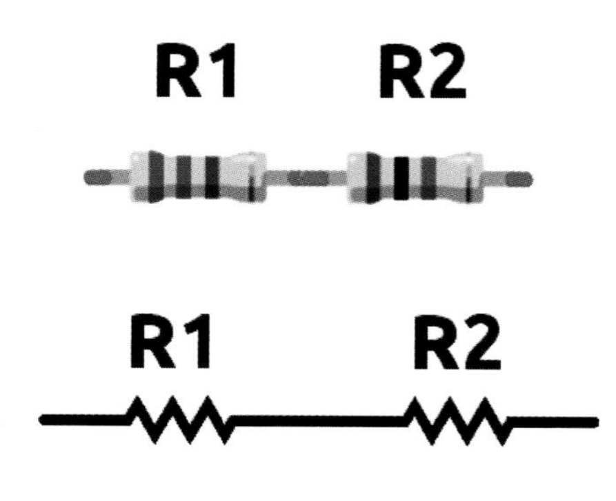

Figure 15-11 *The values of resistances linked in series must be added to one another*

You can link two resistors in a different way too: by coupling their terminals together. In this case the total resistance is calculated with the following formula:

$$R_{total} = (R_1 * R_2) / (R_1 + R_2)$$

It is easy to verify that the resulting resistance decreases. In particular, if the two resistances R_1 and R_2 are equal, the resulting resistance is equal to half of the single resistance's value. We add resistances, but the total resistance decreases. So in Figure 15-12, the resistance is only 500:

```
(1,000 * 1,000) / (1,000 + 1,000) =
1,000,000 / 2,000 = 500
```

How is this possible? Think of the hose analogy: although this arrangement gives you two obstructions, you also have two hoses, so, on the whole, more water flows through.

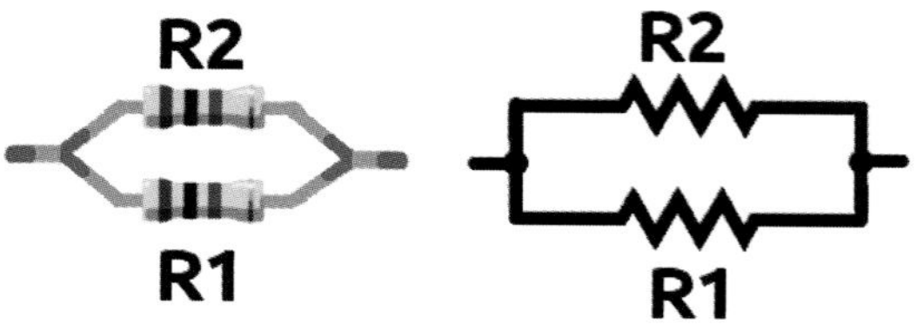

Figure 15-12 *Two parallel resistances*

The resistors used in electronic circuits can vary from tens of ohms up to millions of ohms. Resistors having a higher value obstruct the flowing current more.

When a rope runs through our hands too quickly, we risk getting serious burns from friction. Similarly, when current flows through a resistor, it also produces heat, and excessive heat risks damaging the components of our circuit. Resistors must not be chosen randomly; you need to use the right resistors for each circuit. You'll see how to choose the right ones in "Ohm's Law" on page 139.

Trimmers and potentiometers

There are *variable resistors* whose value you can set by turning a control knob, just like you do with water taps, in which a valve widens or shrinks the diameter of an obstruction and lets more or less water flow through. There are two types, shown in Figure 15-13:

- You will use a *potentiometer* if you have to frequently change the resistance value (for example, when you turn up or down the volume control of an old-fashioned audio system).
- *Trimmers* have no control knob; they are set with a screwdriver and are used when you have to change the value rarely (for example, on light sensors).

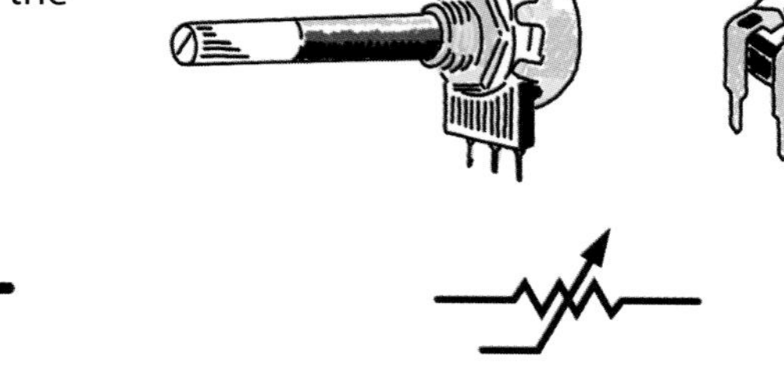

Figure 15-13 *A potentiometer and a trimmer with their electrical symbol*

Both have two or three terminals, or clips, and they can be used in two different ways (see Figure 15-14):

- As a *variable resistor*: you connect them so that only two pins are used.
- As a *voltage divider*: you connect all three pins, and the voltage on the second pin varies based on where you've positioned the knob or the screw. See "Using a Voltage Divider" on page 133 for more details.

Using a Voltage Divider

Using a voltage divider involves different math than the earlier calculation where you read the total value of resistors in series. With a voltage divider, you calculate the voltage produced by looking at the input voltage, the value of the resistor closest to the positive terminal (R_1), and the value of the resistor closest to the negative terminal (R_2). When you turn the knob or screw such that the resistance of R_1 increases, R_2 decreases accordingly. The terminal with the input voltage is labeled V_{cc} in Figure 15-14, and the output voltage as V_a. The terminal marked GND is connected to ground (negative):

$$V_a = V_{cc} * (R_2 / (R_1 + R_2)$$

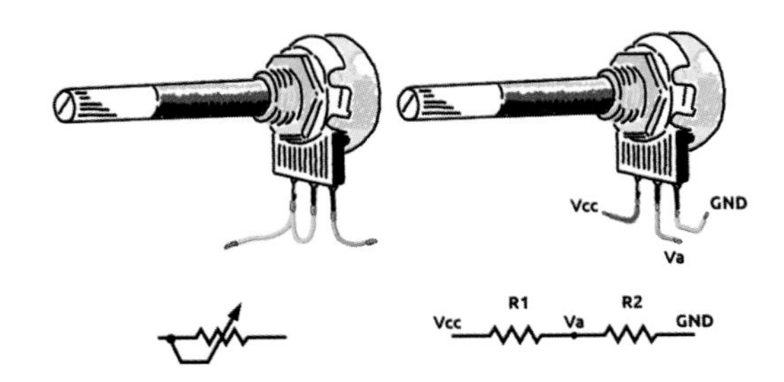

Figure 15-14 *The two possible ways of using a potentiometer*

Capacitors

You can think of a *capacitor* as similar to a cistern in which you store small quantities of water. The capacitor's capacity is measured in *farads* (F). Typical values range from microfarads (µF, 0.000001F farad), for the components used in chargers or audio amplifiers, to picofarads (pF, 0.000000000001F farad), for radio or computer components.

If you were to create a capacitor you'd make a sort of a sandwich with two layers of conductive material padded with an insulating layer (or with air) so that the current has an incredibly difficult time passing through. As insulating material you'd use ceramic, plastic, paper, liquids, and certain metals. For this reason it is possible to find capacitors in different shapes and sizes, as shown in Figure 15-15. Because of the materials used for their manufacturing, some kinds of capacitors, like electrolytic capacitors, also have a polarity; one side is positive, while the other side is negative.

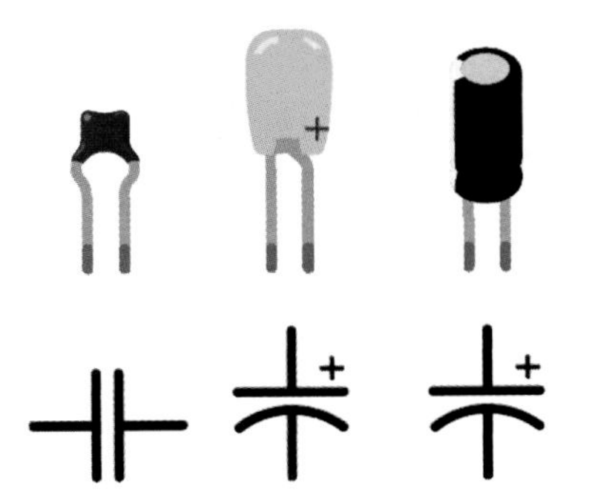

Figure 15-15 *Some capacitors and their corresponding symbols*

What would happen if you tried to add a capacitor of some hundreds of µF to the Hello World circuit? Figure 15-16 shows what it would look like.

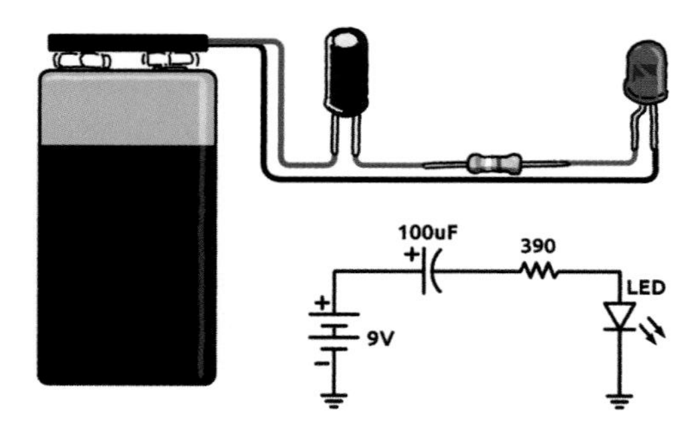

Figure 15-16 *Modifying the Hello World circuit by adding a capacitor*

The LED lights up, but only for a fraction of a second, and it turns off again straightaway. How is this possible? Capacitors permit only alternating current to pass through. If you disconnected the capacitor and you connected it directly, but inverted, to the LED pins (Figure 15-17), the LED would light up again for a short moment, thus stealing away the energy stored by the capacitor.

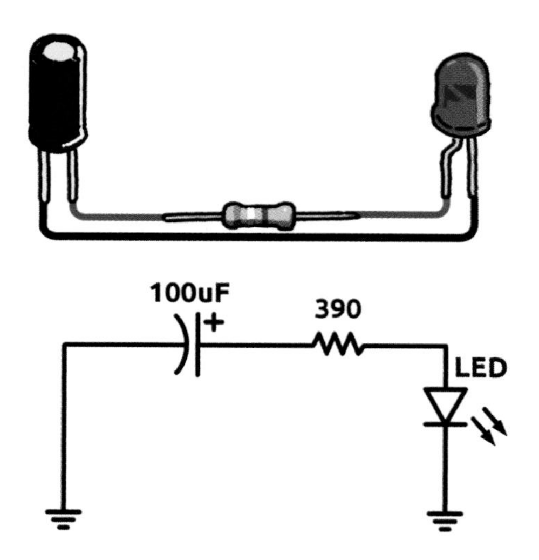

Figure 15-17 *By disconnecting the battery and connecting the capacitor, inverted, you discharge it*

Diodes

A wire and a resistor both let current flow in either direction. It doesn't matter how you orient these components in your circuit. As we mentioned earlier, a *diode* (Figure 15-18) is a component that lets current flow easily in only *one* direction. It is used to prevent current from flowing in the wrong direction to protect certain components that would be damaged if we allowed current to flow through them. In our comparison with water, a diode is like a nonreturn valve that lets water flow through in one direction only.

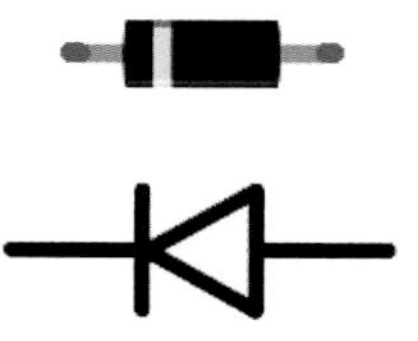

Figure 15-18 *A diode and its symbol*

Buttons and switches

One of the simplest ways to let people interact with an artifact is to make them press a *button*. A button is a switch—a mechanical device that completes a circuit, letting electricity flow, or breaks a circuit, stopping the flow. There are two types of buttons:

- *Normally open buttons*, which (untouched) prevent the current from flowing; when you press them, you close the circuit and the current can flow through.
- *Normally closed buttons*, which (untouched) let the current flow through; when you press them, you open the circuit and stop the current from flowing.

Switches work in a similar way, but stay in the position you put them in. Pushbuttons are commonly *momentary*; their actions are in effect only while you are pressing the button. A switch is a toggle (Figure 15-19)—if you turn a switch off, it stays off until you turn it back on again. Our homes are full of switches: we use them to turn lights on and off.

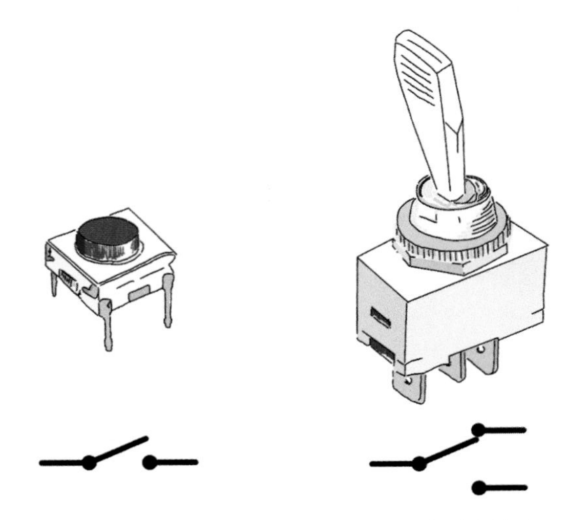

Figure 15-19 *A button and a toggle switch with their corresponding symbols*

Tools

First of all, to work comfortably, you need a well-lit working surface. If you use a table it is best to cover it with a plywood or thick cardboard layer to protect its surface. The tools needed to start building small circuits are few, and they are not expensive. At first a pocket knife, a pair of scissors, and a screwdriver might be more than enough. However, as the circuits get more complex, to do a good job you will need suitable tools, shown in Figure 15-20. The more effective and higher quality they are, the better the result and the lower your strain.

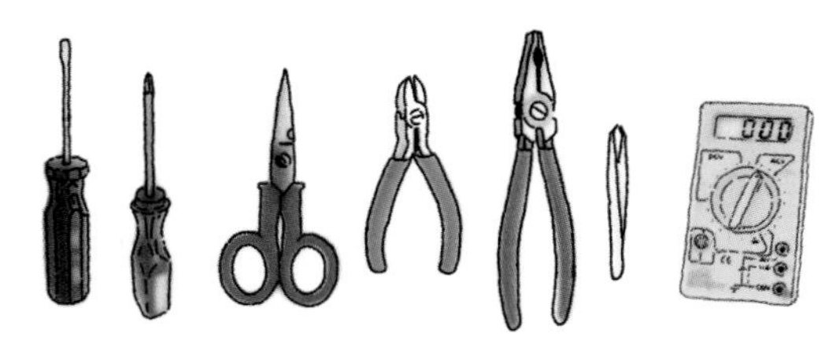

Figure 15-20 *A minimum set of tools*

The basic tools are screwdrivers, tweezers, electrician's scissors, clippers, wire stripper, a multimeter (see "Measurements" on page 137), and, maybe, a magnifying glass. After the very first experiments you will need a soldering iron and solder, a desoldering pump, a sponge for the soldering tip, and finally a "third hand" device to help you hold the components. Figure 15-21 shows these.

Figure 15-21 *The necessary kit for soldering*

Although it is possible to work with batteries, it is always better to ensure a power supply in order to have a stable and reliable supply voltage. Professional labs are equipped with different power supplies in terms of voltage and current, able to support any circuit. For the simplest experiments you can make do with some power supplies found at home, like the charger of an old mobile phone.

Creating a Circuit

Figure 15-22 shows the ways that wires may be represented in a circuit diagram (*schematic*). From left to right, the first two images are used to indicate when wires cross in the diagram but shouldn't be connected together (sometimes, it's not possible to draw the lines any way aside from overlapping). The last image shows wires that must be connected.

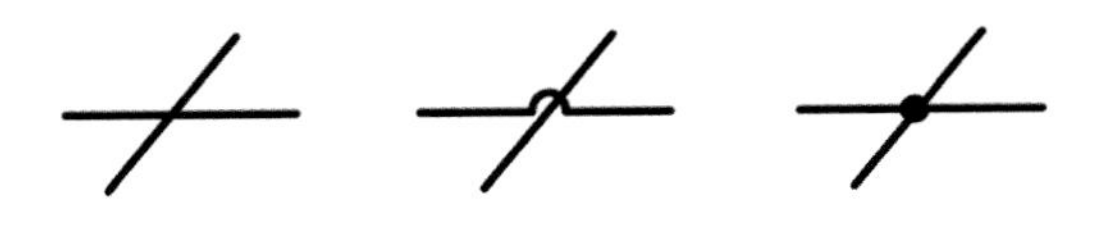

Figure 15-22 *Wires crossing in a circuit: only in the third case are the wires in contact*

The circuit diagram usually has little to do with the final circuit for several reasons: the pins of a component may be in a different order from the symbol, the diagram does not consider the actual dimensions of the components, and so on.

How do you make an electronic circuit? The quickest, simplest, and least risky way is to use a prototyping *breadboard* (Figure 15-23), so called because it looks like a bread cutting board composed of a pierced grid on a tray: when bread is cut, crumbs fall through the holes and are collected in the tray.

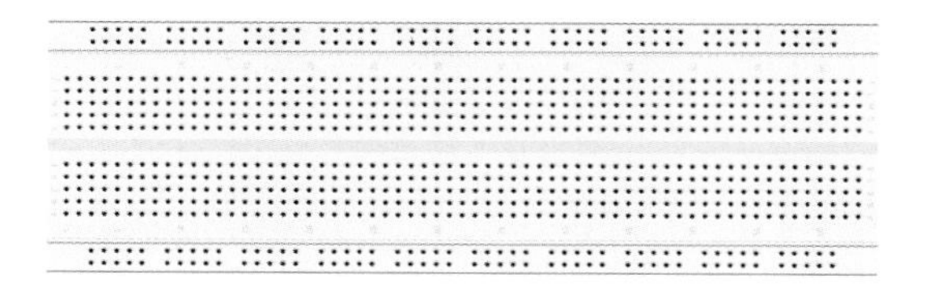

Figure 15-23 *A breadboard*

Using a breadboard is a little like playing with LEGO building blocks: components are inserted into little sockets. Unlike with LEGOs, they are then linked with wire. You can use simple solid-core wires or wires called *jumpers* equipped with handy connectors on their ends (Figure 15-24).

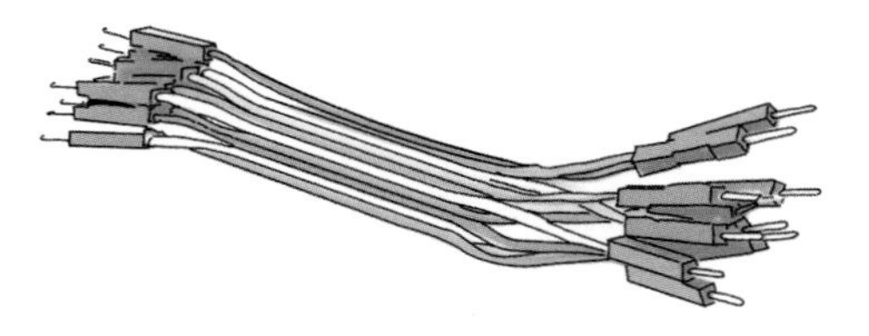

Figure 15-24 *Jumpers for the links on a breadboard*

Breadboards are almost always divided into left and right sectors, with a nonconductive "gutter" running between and separating them. The sockets of each half line are all electrically linked, row by row, as shown in Figures 15-25 and 15-26.

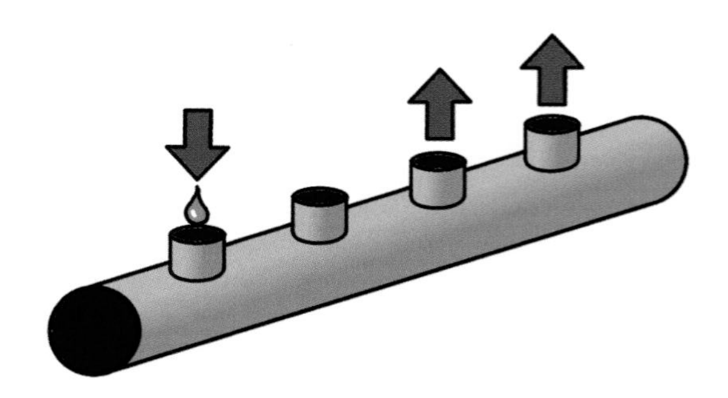

Figure 15-25 *The half lines of a breadboard are like many T-joints linked together*

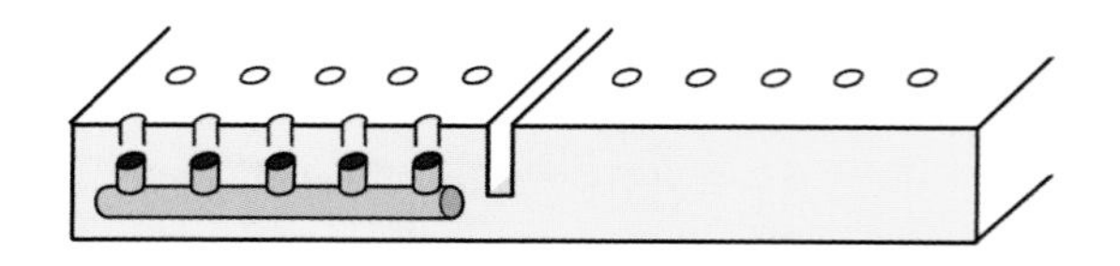

Figure 15-26 *All sockets of a half line are linked to one another*

Breadboards are very practical and quick, but using them for anything other than the absolutely simplest circuits tends to lead to a tangle of wires and unstable situations. Therefore, it is extremely important to try to be tidy and find an optimal arrangement of the components.

Tips on how to arrange the breadboard

The biggest breadboards usually have two pairs of side lines called *rails* used to bring electricity to the board. Rails are more or less at the same distance from all components and prevent tangles from developing. You can place jumper wires across them to share the voltage with both sets of rails, as shown in Figure 15-27.

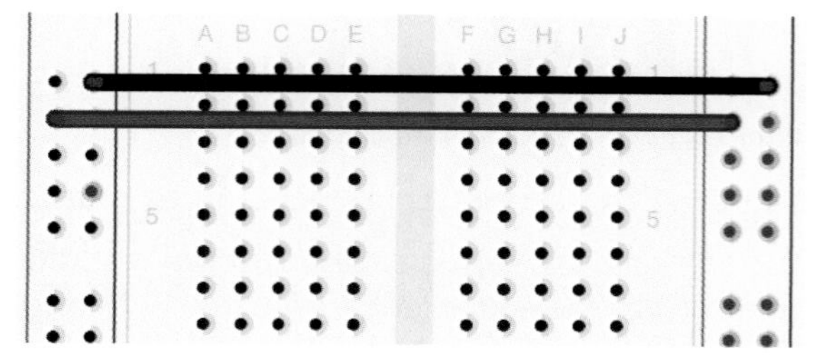

Figure 15-27 *A breadboard with the rails of both sides linked together*

If the circuit requires a double power supply, for example with 5 and 12 volts, it is better to dedicate one rail to 5 V and the other to 12 V. The wires of the negative pole of the power supply (sometimes called "ground" or GND) must always be linked with one another. To minimize electrical interference, add 100 nF capacitors directly onto the rails, as shown in Figure 15-28.

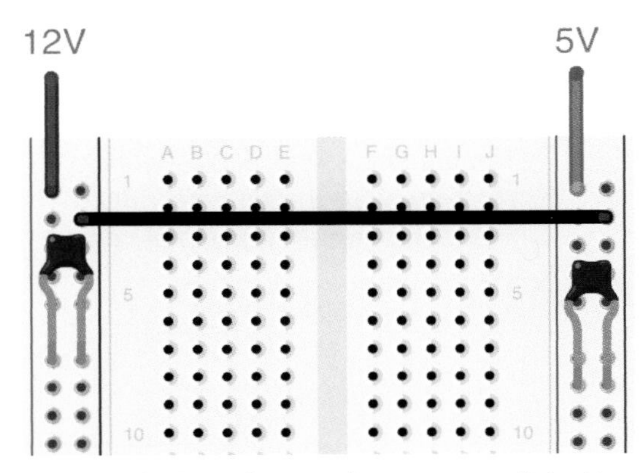

Figure 15-28 *Two different voltages: masses linked to one another, capacitors to diminish disturbances*

On some breadboards, the rails are divided in half, so it is necessary to link them with a small U-shaped piece of wire.

Avoid rings and spirals of wire: if the circuit is particularly delicate or sensitive those loops may behave like the antennas of a radio, which is not always (i.e., hardly ever) what you want, because it can cause unpredictable behavior in your circuit and may emit undesirable radio emissions. Try to use only wires of precisely the right length. Assemble everything with care without overlapping the links.

You can't put all the components in any order or position you like on a breadboard. They must connect to each other the way the circuit design commands. You can adapt potentiometers, speakers, relays, and charging plugs by soldering or linking wires to their terminals.

If we want a more solid and resistant circuit we can use a soldering iron to fix the components to some pierced plates called *matrix boards*, or even create a *printed circuit board* (PCB). Figures 15-29 and 15-30 show both types.

Figure 15-29 *A prototype*

Figure 15-30 *The ABNormal board on a PCB*

Measurements

Earlier, you saw that current and voltage are measured with similar-looking tools. Someone had the great idea to create a tool able to measure both quantities, as well as other electric quantities: the *multimeter*. Both digital and analog multimeters are available: the analog ones, which show values via needles, are still very popular because they measure the signals directly. Digital multimeters, on the other hand, indicate measurements with numbers on an LCD display; fluctuation in the signal shows up as wildly varying values on the display screen.

A multimeter is equipped with a pair of inputs, usually a red and a black wire with pen-like probes attached to the ends. The black probe reads the common or ground signal (represented as COM or GND). The red probe reads the quantity we want to measure—voltage, resistance, or amperage—which we usually select by turning a control knob, as shown in Figure 15-31.

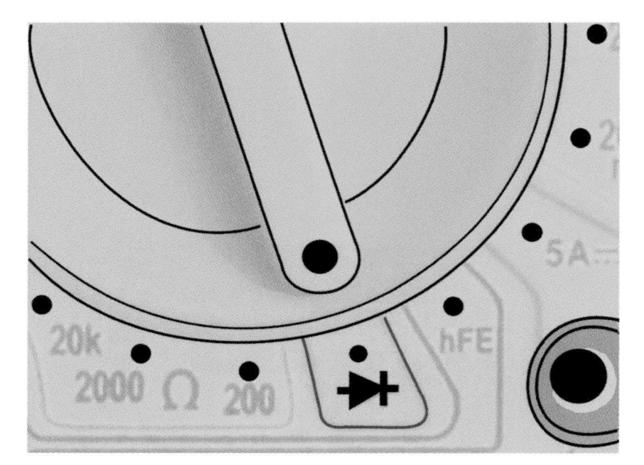

Figure 15-31 *A multimeter control knob set to test a diode*

Voltage is a simple quantity to measure. Make sure the probes are inserted in the correct socket on the multimeter: the black probe in the COM socket and the red probe in the V or DCV socket. Turn the control knob to an appropriate DCV position; the voltage you measure must be lower than the labeled setting on the multimeter. For example, if you're measuring something around 7 volts, set the control knob to 10 or 20 volts. If you are unsure, start with the highest setting, and work your way down. To measure the voltage between two points of the circuit, simply touch the probes to exposed conductors (either exposed wire, or exposed component pins) on the part of the circuit you are interested in, as shown in Figure 15-32.

AC/DC

You need to know whether you're measuring AC or DC voltage. The probe wire goes into a different jack on the multimeter when measuring AC, and most AC voltages you'll encounter are life-threatening. You can seriously damage your multimeter (not to mention yourself) if you mess this up.

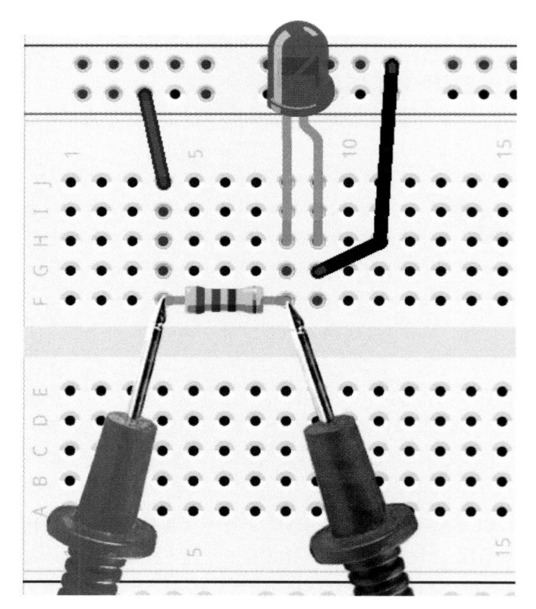

Figure 15-32 *How to measure the voltage at the ends of a resistor*

You can also measure the amperage flowing through part of a circuit. As with measuring the voltage, set the control knob to the appropriate setting for the current you're measuring. Unlike with voltage, you need to "break" the circuit where you want to measure and place the meter across the break, as shown in Figure 15-33. The value registered by the multimeter is the intensity of the current that flows through this point in the circuit.

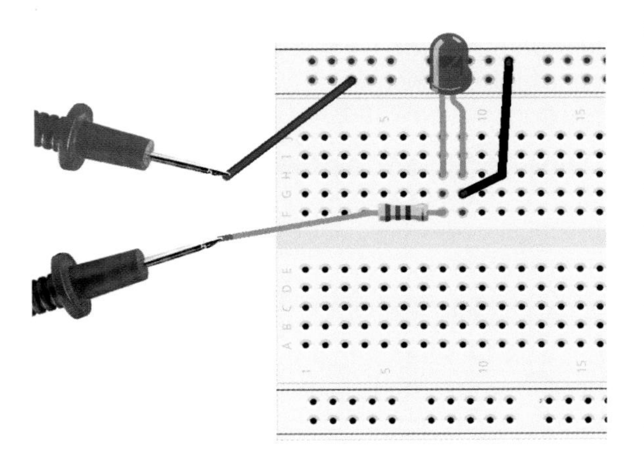

Figure 15-33 *How to measure the current flowing through a resistor*

Ohm's Law

Many components you use in your circuits work only if you supply the right voltage. Continuing the metaphor that voltage is much like water pressure, it's a bit like making the wheel of a mill turn. If you placed the mill wheel under Niagara Falls, it would be destroyed immediately; the pressure would be too high for it. If you tried to make the mill wheel turn with the stream from a kitchen tap, it wouldn't move at all.

Current, voltage, and resistance are not independent quantities, but, as we said before, they are linked by Ohm's law:

```
V = R * I
```

V is voltage, I is current intensity, and R is resistance. This law is extremely important because it allows you to make the necessary calculations to protect your circuits. You will know what voltage and what resistors you have to supply your circuits with so that everything will work perfectly and nothing is damaged.

Let's go back to the blinking LED version of Hello World. Have we chosen the right resistor?

To work correctly, the LED requires a precise voltage and a certain quantity of current, both of which you can obtain from a datasheet. Usually, the charging supply voltage depends on the color and the type of LED, while the working current ranges from 10 to 20 mA. See Table 15-5 for some typical values.

Table 15-5 *Typical values for the supply charging voltages of LEDs*

LED color	Supply charging voltage (V)
Red	1.8
Yellow	1.9
Green	2.0
Orange	2.0
Blue/white	3.0
Blue	3.5

Suppose you are using a 9 volt battery and an orange LED: if you linked them directly, the LED would get burnt (just like the wheel of a small mill would break under Niagara Falls). If you need 2 volts for the LED, a 9 volt battery supplies 7 volts too many. What do we do?

This is why we need a resistor: we are going to use it to divide the waterfall into two smaller falls so that the second one has a suitable height for our mill (Figure 15-34). The waterfalls are shown as arrows. The arrows' length is equal to the water drop. The first arrow, on the left, represents the battery and is 9 units long. We then draw two more arrows on the left (of 7 and 2 volts). The loop must always be closed, and we must have a balance between what we have on the right and on the left: this is just a graphic representation of *Kirchhoff's current law*. What resistance do we need in order to get the 7 volt voltage needed to avoid breaking our LED?

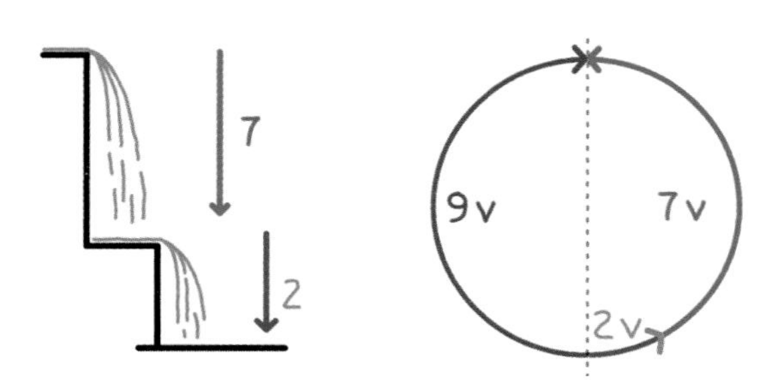

Figure 15-34 *Split the waterfall into two small waterfalls*

With some mathematics you can write Ohm's law as follows:

```
R = V/I
```

We have said that you need to divert 7 volts. From the LED datasheet we know that 20 mA—that is, 0.020 A—must flow through the LED. So, if we replace the values in the equation we get:

```
R = 7V / 0.020 A = 350 Ω
```

We will not find a 350 ohm resistor on the market, because they are manufactured with standard values, but we can come very close: the nearest standard is 348 ohms! While this will be acceptable for most purposes, it is always better to reduce the current a bit too much than to put too much current through a device. When faced with this situation, while we *can* use a 348 ohm resistor, we should probably use the next highest one, which would be a 360 ohm resistor, or even a 390.

Ohm's law also explains why we have to avoid short circuits: in a short circuit the resistance is practically zero. Let's write Ohm's law again, this time as follows:

```
I = V / R
```

It is easy to see that in a short circuit with practically zero resistance, we are dividing the voltage by a very small number, so the current will tend toward a very high value. Our power source will do its best to try to supply it, but at a certain point it will reach its limit and something will break. Even with "just" a 12V car battery, a big screwdriver placed across the terminals can get (partly) disintegrated and harm the unlucky human who put it there!

Arduino

16

Once upon a time, you could be considered an electrical wizard simply by creating a pushbutton electric doorbell for your house. As the field of electronics grew more complicated, hobby electronics also grew complicated. Soon hobbyists were making their own tube radios, then transistor radios, then their own televisions. In 1975, the field reached a peak of complexity, when hobbyists could even build their own desktop computer, the Altair. But that high point also marked the beginning of a serious dropoff in home hobby electronics. You didn't need to know much electrical theory to wire up a doorbell, but building a computer required specialized knowledge in mathematics and physics. This was a high barrier that beginners had to overcome, and the learning curve was substantial, especially for people without specific technical competencies. Then, in the mid-2000s, things started to turn around...

What Is Arduino?

Arduino (Figure 16-1) is a sort of small computer—simple, powerful, affordable, and versatile —that can interact with the surrounding physical world. You can connect it to sensors (devices that give you information) and actuators (devices able to perform actions), and program it to follow rules and instructions that you set. For example, you can use Arduino to control a greenhouse: when the humidity level changes, Arduino can start or stop the ventilators to keep optimum conditions for your plants.

Arduino grew out of work done at the Interaction Design Institute Ivrea in Ivrea, Italy. Students there were building small interactive devices using a microcontroller called a Basic Stamp, which cost about $100. Eager to make interactive computing more affordable, educators and engineers Massimo Banzi, David Cuartielles, Gianluca Martino, and David Mellis created the open source Arduino microcontroller, which retails for about $30. Shortly after the creation of Arduino, Tom Igoe, a professor at the Interactive Telecommunications Program at New York University, joined the team.

Figure 16-1 *Some versions of Arduino, including the special edition for the Maker Faire*

To program Arduino's behavior you need to write software using the Arduino integrated development environment, or IDE. You can download this for free from the official website (*http://arduino.cc*).

Beside interfacing with other physical components, Arduino can communicate with computer-run software, such as *Processing* and *Python*; you'll see some examples of this kind of interaction in Chapter 18 and Chapter 19.

The Arduino board and its programming environment constitute a *prototyping platform*: you can plan and design around the Arduino, then decide to keep Arduino as the core of your project or go with a custom designed solution. After all, an Arduino can do wonderful things, but if you find that you're only using it to switch a light on, there are cheaper solutions.

One of the best things about Arduino is that it is open source: all plans and instructions to make it are publicly available, so anyone could assemble his own board, or even improve it and adapt it to his needs. Another great advantage is that you can use the Arduino development environment with the Big Three operating systems, Windows, Macintosh, and Linux, making Arduino available to as many people as possible.

The Software Structure

Arduino's microcontroller—the "brains" of the board—is rather small as modern-day computers go. It cannot handle large programs like word processors or web browsers. The programs you write—called *sketches* in Arduino parlance—will at most be a few screens long. The language in which you will write these sketches is a simplified version of the more complex C/C++, very similar to Processing, which itself is a simplified version of Java.

Sketches are divided in two separate sections: *setup* and *loop*. The *setup* runs once and handles the preparation for the program. For example, it might tell Arduino that the wire attached to pin 9 is connected to a sensor, that the Arduino should output data to the serial port at a certain rate, or anything else the Arduino needs to know before it gets to work. The *loop* comprises all the actions Arduino will carry out over and over and over, as long as it is supplied with power (for example, "Check the value detected by a sensor. If it is over a certain threshold, switch a light on. If it is under a certain threshold, turn the light off. Wait 10 seconds and start again.").

The sketch is stored in what is called *nonvolatile* memory in the Arduino, which means that Arduino will remember it even after the power is turned off. When you restore power the next day, the next month, even the next year, Arduino will start carrying out its sketch: first the setup, once only, and then the loop, indefinitely.

The Simplest Sketch

The simplest sketch (Example 16-1) doesn't do anything at all. You can't get any easier than this.

Example 16-1 *An empty sketch*

```
// lines starting with // are comments
void setup()
{
  // the initialization code goes here
  // it is carried out only once
}
void loop()
{
  // the main code goes here
  // it is carried out indefinitely
}
```

Not very interesting, is it? Notice two things: first, whatever follows the characters // (double slash) is a comment and will be ignored; second, all `setup` and `loop` sections are marked by curly brackets: you'll see that this is a way to group blocks of instructions.

Luckily, the development environment has many premade examples that show the various behaviors we can set up. Have a look at the *Blink* example (click File→Examples→01.Basics→Blink to load it, as shown in Figure 16-2).

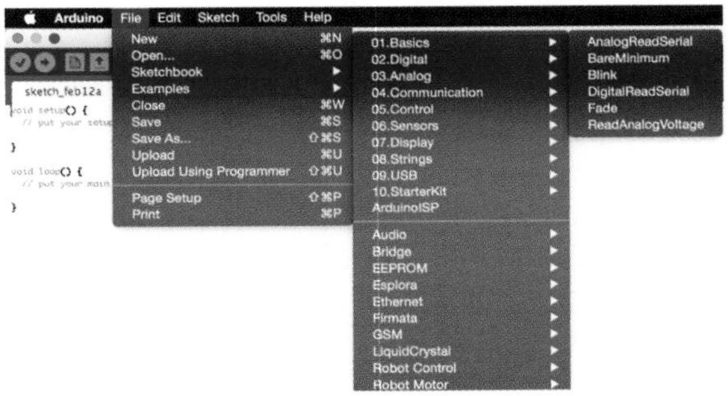

Figure 16-2 *Where to find the "Blink" sketch*

This sketch makes an LED on your board blink (see Figure 16-3); we'll show a slightly modified version with its comments in Example 16-2.

Example 16-2 *Blinking an LED*

```
void setup()
{
 // actuator on pin 13 (specifically, an LED) we want to operate
 pinMode(13, OUTPUT);
}
void loop()
{
  // switch on the LED
  digitalWrite(13, HIGH);
  // wait one second--one thousand milliseconds
  delay(1000);
  // switch off the LED
  digitalWrite(13, LOW);
  // wait one second
  delay(1000);
}
```

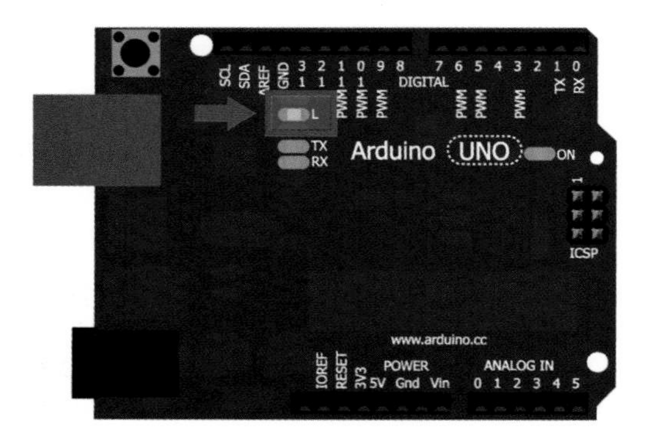

Figure 16-3 *The LED mounted on the board is connected to pin 13*

Let's look at the code. The first thing you probably noticed is that every instruction that is not a comment ends in a semicolon (;). This is because Arduino expects to know where every instruction ends and the next one begins; thus, each statement must be separated with a semicolon. Without the semicolon, the Arduino will refuse to try to make sense of the rest of the program.

As if this weren't enough, the Arduino is so meticulous that it recognizes the difference between upper- and lowercase. The command `DigitalWrite` is distinctly different from `digitalWrite` (the correct form), or even `digital write`.

How to Upload a Sketch in Arduino

So, you've got the Blink sketch in your IDE. How do you get it into the Arduino hardware? Before you upload a program in Arduino, it needs to be *verified*; that is, the development environment has to translate (compile) what we have written into a long series of 0s and 1s (difficult for humans to understand, yet perfectly clear for the hardware) and tell us whether it encountered an error doing so.

Before you verify the sketch, connect your Arduino to your computer using the appropriate USB cable (for example, with the Uno, it's a USB A to B cable; with the Leonardo, a USB Micro cable). Next, click the Tools menu, click Port, and then select the port with the name of your Arduino board next to it (if there's more than one listing for your Arduino, choose either one). Figure 16-4 shows this. The list may be quite long. After you've done that, click Tools→Board and choose your board type (such as Arduino Uno) from the menu. This configures the Arduino IDE to be able to connect to the Arduino (not needed until the next step) and also to use the verification settings appropriate for your model of Arduino.

Figure 16-4 *Choosing the right port*

Now you're ready to verify the sketch: click the Verify button as shown in Figure 16-5.

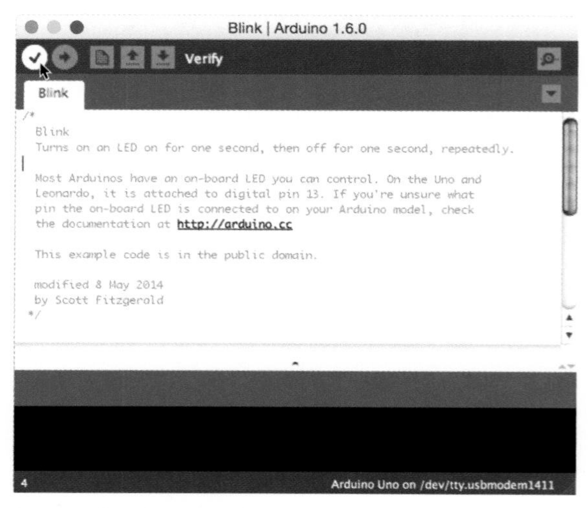

Figure 16-5 *Verifying a sketch*

Once you've verified your sketch, you can upload it to your Arduino. Click the Upload button, and your compiled sketch gets sent to the actual hardware, as shown in Figure 16-6.

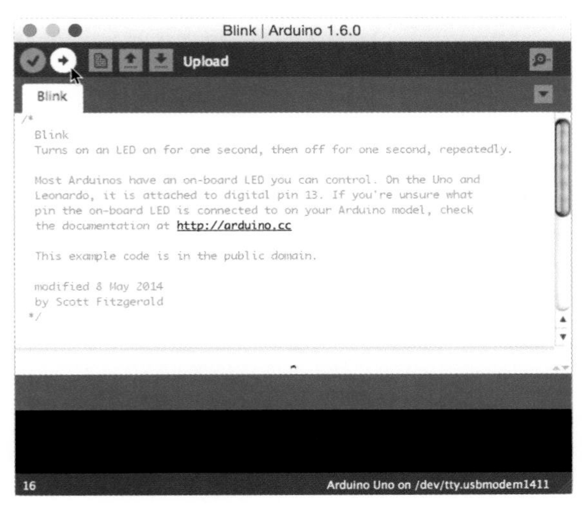

Figure 16-6 *Uploading a sketch*

If everything goes well, you'll see the LEDs shown in Figure 16-7 flash, indicating that the computer is transferring your compiled instructions to Arduino. After a few seconds you'll see the LED on Arduino blink steadily.

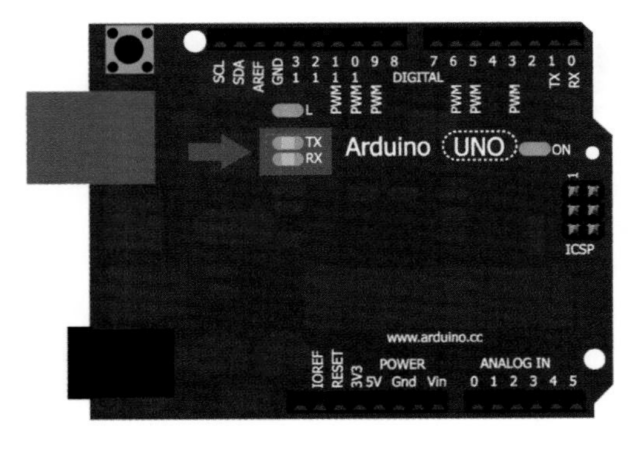

Figure 16-7 *The TX and RX LEDs show your upload in action*

In the next section, we'll go into some more detail:

- How to tell Arduino where you want to connect sensors and actuators
- How to read the sensor values and tell the actuators what to do

Interacting with the Physical World

Let's introduce the first simple commands that we can use to give our creations life.

Defining a pin's behavior

On an Arduino board there are several pins; in Figure 16-8 you'll see some of them highlighted, numbered from 0 to 13. These pins are all digital, only able to process 0 and 1, all or nothing. If they were water taps they could only be either fully open or fully closed.

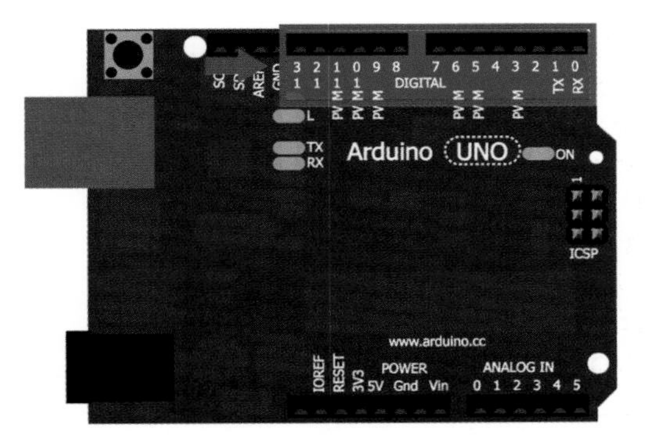

Figure 16-8 *Digital pins of Arduino*

One of the advantages of a prototyping platform like Arduino is the ability to change the digital pins' behavior and decide whether we want to use them as *inputs* or *outputs*. The instructions you need to put into the setup block for either are very easy. For example, to set pin number 13 as an output, simply add the following code:

```
pinMode(13, OUTPUT);
```

Whereas if we want pin number 7 to act as an input, the instruction is:

```
pinMode(7, INPUT);
```

Basically, the `pinMode` command accepts two parameters: the first one determines which pin it will operate on, and the second one tells whether the pin must be an input (`INPUT`) or an output (`OUTPUT`). If you declare a pin to be an input and then try to output data to it, or the other way around, the Arduino or the attached electronic might get damaged.

Shall We Switch It On?

If you have identified a pin as an output, you can "switch it on" or "switch it off" using the command `digitalWrite`. This function accepts two parameters: the first one is the pin that we want to operate on, and the second one represents what you want it to do (i.e., whether or not any current will be let out from the pin). As mentioned earlier, with digital pins, you can only switch them on or off—the tap is fully open or fully closed, no midway.

To "switch on" pin 13, you'd write:

```
digitalWrite(13, HIGH);
```

To "switch it off," you'd write:

```
digitalWrite(13, LOW);
```

You see both these instructions in the Blink sketch. The LED on the board, connected to pin 13, switches on and off when the value of the second parameter is `HIGH` or `LOW`, respectively.

Not So Fast...

Sometimes you might want to slow the action down. You can add a pause into the execution of the instructions that Arduino follows. To do this, you can use the function `delay`, which needs a parameter indicating the number of milliseconds that the Arduino has to wait before carrying out the next instruction. For example, to tell Arduino to stop for one second, you would write:

```
delay(1000);
```

You need to write `1000` instead of `1` because 1,000 milliseconds equals 1 second. As usual, don't forget the semicolon at the end. For a 2.5-second pause, you'd write:

```
delay(2500);
```

Edit the Blink sketch to change the length of the delay, then verify and upload it; you'll see the difference straightaway when the LED blinks at a different rate than before.

Pardon Me, You Were Saying?

You now know how to have Arduino carry out an action, even if it is just switching on an LED. Now let's teach this LED to switch on only in a certain condition: when you press a button. How can you do that? Let's start by assembling the circuit. You'll need the following to do this (everything listed here, with the exception of the Arduino Uno, is included in Maker Shed's Mintronics Survival Pack (*http://www.makershed.com/products/mintronics-survival-pack*)):

- One LED
- One 220 ohm resistor
- One 10 kilohm (aka 10K) resistor
- A pushbutton
- Assorted jumper wires
- A mini breadboard

Start with the LED. Rather than using the onboard one (Figure 16-3), which is small, we're using an external one. Look at Figure 16-9 to see how to set it up. Note that you're connecting it across the gap in the breadboard, so you need wires both from GND and pin 13 to the breadboard, and another wire to bridge the gap to the LED's negative lead. For the positive lead of the LED, you're bridging the gap with a 220 ohm resistor to protect the LED and the Arduino from overcurrent. See "Resistors" on page 131 for a refresher on working with resistors.

For the button, you also need to add a resistor, to protect the Arduino's input pin and to make the input circuit more reliable. A 10 kilohm (10K) resistor will do. Take the resistor and use it to connect the column that the GND wire is in to the column that the button is in. Then, connect a wire from pin 7 to where the resistor meets the button, as shown in Figure 16-9. Finally, connect the 5V pin to the button's other lead as shown.

To set pin 7 to input mode, you already know you have to write:

```
pinMode(7, INPUT);
```

But how do you know if the button is pressed or not? The function to find out if voltage on the pin is `HIGH` (pressed) or `LOW` (not pressed) is `digitalRead`. As with other functions, you have to specify which pin you want to read, in this case number 7:

```
digitalRead(7);
```

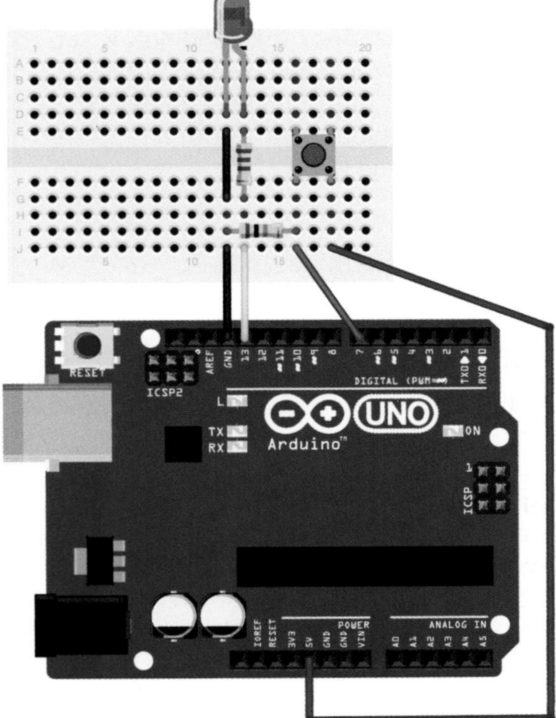

Figure 16-9 *Using a button to switch an LED on*

The `digitalRead` command is a little different from the functions you've seen so far, in that it returns a value: `HIGH` if there's a voltage on the pin and `LOW` if not.

Where Do You Store Your Data?

What should we do with the value that the `digitalRead` command returns? You can use it to set the value of a *variable*, which is kind of like a drawer with a name tag on it where you put a value that you can check or edit any time you want. Let's create a variable named `read` and give it the value returned by `digitalRead`. To do that, use the symbol `=` (the assignment operator):

```
int read = digitalRead(7);
```

The `int` in front of the name of the variable is a special term used to define what kind of data you'll put in your "drawer." In this case, you're stating that the variable `read` will have an integer value (between –32,768 and 32,767). But... didn't we say that `digitalRead` can return only `HIGH` or `LOW`? Right! These two values are *constants*, which you use to remember the meaning of the associated values. In the Arduino language there are many predefined constants; for example, the values `INPUT` and `OUTPUT`, which you used to define the pin behavior, are constants. Arduino's built-in constants are special constants called *preprocessor macros*, and we suggest you don't define your own macros (for consistency with the Arduino programming style). If needed, you could write your own constants using a different method. You just need to put the term `const` in front of the statement you have already seen for variables:

```
const int RINGS_FOR_THE_ELVEN_KINGS = 3;
```

Conventionally, only uppercase is used for constant names. This is in contrast to variables, whose value you can change any time you want. Once you have defined a constant it remains, indeed, constant and unchanging.

You can choose the name of your variables and constants almost as you wish, as long as you stick with the characters A–Z, 0–9, and "_" (underscore), and avoid conflicts with other, already defined names. In order to improve the understanding of sketches you should choose meaningful names; while strange names such as AUF2RTX are valid, they're not very meaningful, so it's best to avoid them.

What should we write in our sketch? In the setup, you need to state that pin 7 is an input and 13 an output. You already know that you can write:

```
void setup()
{
   pinMode(13, OUTPUT);
   pinMode(7, INPUT);
}
```

However, with this approach you would have to remember throughout the whole sketch that the LED is on pin 13 and the button is on pin 7. Thanks to the variables, you could write this instead:

```
int led = 13;
int button = 7;
void setup()
{
   pinMode(led, OUTPUT);
   pinMode(button, INPUT);
}
```

With this change you will be able to work out whether the button is pressed with a simple `digitalRead(button);` and switch on the LED by writing `digitalWrit`his is because if you declared the variables within the setup block, you could use them only in that block. This principle applies for all blocks, and is known as *visibility of variables* or *scope*.

Normally, a computer executes all the instructions of the program, from start to finish, only once for the setup and continuously for the loop. Sometimes you want to modify this flow; for instance, you may want to see one action repeated several times or execute some of the instructions only upon certain conditions.

Only When I Say Go...

As we were saying, you want the light to switch on only if you are pressing the button. For this, you need to use a *control structure*. Let's think about how that would work, using *pseudocode* (human-readable instructions that are structured like real computer code):

```
if this happens:
  do that
```

The keyword you'd use to take action under a certain condition is `if`:

```
if (condition)
{
  // list of commands to execute
  // there can be many of them!
}
```

Conditions can be based on the reading you get from a sensor, the number of times you have already followed some instruction, the value of a variable, and much else. Notice that the actual condition to be checked is in parentheses, and that there is no semicolon at the end of the line. After the condition is set, the action to be taken when the condition is true is set off by curly braces ({ and }). In this case, you want to verify whether the button is pressed or not, so you could write:

```
if (digitalRead(button) == HIGH)
{
  // The code to switch on the LED
  // would go here.
}
```

When Equals Is Not Equals

You might have noticed how, when you need to verify whether two quantities are equal, you use the == sign (double equals signs). You need to be very careful, because a single equals sign is used to give a variable a value!

```
int tester;
void setup() {
  tester = 0; // set tester equal to 5
}
void loop() {
  // Some code that affects tester goes
  // here.

  if (tester == 5) // checks if tester
                   // is equal to 5
  {
    // do something
  }
}
```

If you were to write `if (tester = 5)` instead of using ==, then `tester` would be set to 5 and the condition would always be true!

In the sketch, if the button is not pressed, you want the light to stay off. To introduce an alternative behavior to be used when the condition is not verified, use the else block:

```
if (digitalRead(button) == HIGH)
{
  // switch on the LED
}
else
{
  // switch off the LED
}
```

The else block is not mandatory; many if statements you write will not have one.

We are getting closer to the goal...you have seen how to switch an LED on and off, so you know that you only need to write:

```
if (digitalRead(button) == HIGH)
{
   digitalWrite(led, HIGH);
}
else
{
   digitalWrite(led, LOW);
}
```

Because you want the Arduino to keep repeating this behavior, this part of the code will be put into the loop block. The complete code for this example is shown in Example 16-3.

Example 16-3 *The LED switches on only when you press the button*

```
int led = 13;
int button = 7;
void setup()
{
   pinMode(led, OUTPUT);
   pinMode(button, INPUT);
}
void loop()
{
   if (digitalRead(button) == HIGH)
   {
      digitalWrite(led, HIGH);
   }
   else
   {
      digitalWrite(led, LOW);
   }
}
```

Now all that's left to do is to click the Verify button, then use the Upload button to send the program to the Arduino and test the outcome of your efforts.

Congratulate yourself: you have made your first interactive prototype!

...Even Two, Three Times!

Suppose you want some instruction to be repeated a number of times. For example, you can try to have the LED flash three times each time you press the button. It will stay on for one second, and wait another second before blinking again. It is interesting to see how, in order to change the behavior of this prototype, you don't need to modify anything on the *circuit*—

you only need to edit the *sketch*. Isn't this great? With what you have learned so far, you are already able to make the necessary changes shown in Example 16-4.

Example 16-4 *How to repeat instructions*

```
int led = 13;
int button = 7;
void setup()
{
   pinMode(led, OUTPUT);
   pinMode(button, INPUT);
}

void loop()
{
   if (digitalRead(button) == HIGH)
   {
      // first time
      digitalWrite(led, HIGH);
      delay(1000);
      digitalWrite(led, LOW);
      delay(1000);
      // second time
      digitalWrite(led, HIGH);
      delay(1000);
      digitalWrite(led, LOW);
      delay(1000);
      // third time
      digitalWrite(led, HIGH);
      delay(1000);
      digitalWrite(led, LOW);
      delay(1000);
   }
   else
   {
      digitalWrite(led, LOW);
   }
}
```

Verify and upload the sketch to check that everything works as you expect. Done? Great. What if, instead of having the LED flashing 3 times, you wanted it to blink 5 times? Or 10? Or 20? You could, of course, copy the lines that correspond to the blinking as many times as you need, but it wouldn't be very practical. Moreover, what if you wanted the LED to blink for a number of times that was contingent upon something else? You need to place the blinking routine into a structure that will repeat the blinking a variable number of times.

In these cases, *loops* help. To execute a group of instructions a fixed number of times, you use a special keyword: `for`. You may remember seeing `for` loops from when we were working with OpenSCAD in Chapter 11. Arduino's loops are similar in structure, as shown in Example 16-5.

Example 16-5 *Syntax of the for loop*

```
int numberOfRepetitions = 10;
for (int i = 0; i < numberOfRepetitions; i++)
{
  // the list of commands to be repeated goes here
}
```

You put three pieces of information in the heading of a `for` loop:

- A counter—that is, a variable that keeps track of how many times you have already executed the instruction block inside the loop. You also have to specify the initial value to start counting with. In this example, you can write this as `int i = 0;`.
- A condition, to specify when the loop can stop. In this case, the condition is `i< numberOfRepetitions`, where `numberOfRepetitions` is a variable that has already been given the value 10. The loop repeats until the condition becomes true.
- A way to increase the counter. In this example we have used `i++`, which is an abbreviated way to say `i = i + 1`.

The three parts of a `for` loop are placed within parentheses and separated by semicolons. Basically, the loop works as follows:

1. Set the initial value of the counter.
2. Verify the value of the counter.
3. If the counter value meets the condition (i.e., the counter value is lower than the comparison value), the instructions block is executed; if not, the cycle ends.
4. Increase the value of the counter and go back to step 2.

Typically, counters use variables with very short names, such as `i`, `j`, or `k`. Nothing prevents you from choosing longer names, but it is not very practical and doesn't add any value. In this example, in order to have the LED blink 10 times, you'd use the modified `loop` function shown in Example 16-6.

Example 16-6 *Repeating instructions without being masochists*

```
void loop()
{
  // check if the button is pressed
  if (digitalRead(button) == HIGH)
  {
     for (int i = 0; i < 10; i++) // Begin the for loop
     {
       digitalWrite(led, HIGH);
       delay(1000);
       digitalWrite(led, LOW);
       delay(1000);
     } // Close the for loop
  } // close the if block
```

```
    else
    { // the button is not pressed
      digitalWrite(led, LOW);
    } // close the else block
  } // close the loop
```

When the cycle is finished, the loop block starts again, and so on until you switch the power off. For practice, try to change the code to make the LED blink 20 times with an interval of one-tenth of a second.

You can count in other ways; for example, you can count backward, decreasing the value of the counter by using `i--`, which is an abbreviated way of saying `i = i - 1`. You could even skip some count values, for example by counting by two. The increment statement in the loop, in this case, would be `i = i + 2`.

Beyond Digital

Until now, you have used only digital values: on or off, open or closed, 1 or 0. However, the real world is not only black or white; it is all the shades between. Analog sensors (like the potentiometers you saw in "Trimmers and potentiometers" on page 132) provide results in those shades between black and white. There are many other sensors that can read values along this range of possibilities and help Arduino make decisions based on them. For a book full of sensor tutorials and projects, see *Make: Sensors* by Tero Karvinen, Kimmo Karvinen, and Ville Valtokari.

I fully half-agree

The pins you saw in the previous section can read only digital signals. Luckily (or more accurately, thanks to a specific and sensible design choice), on Arduino there are also pins capable of reading analog values: all pins from A0 to A5, shown in Figure 16-10.

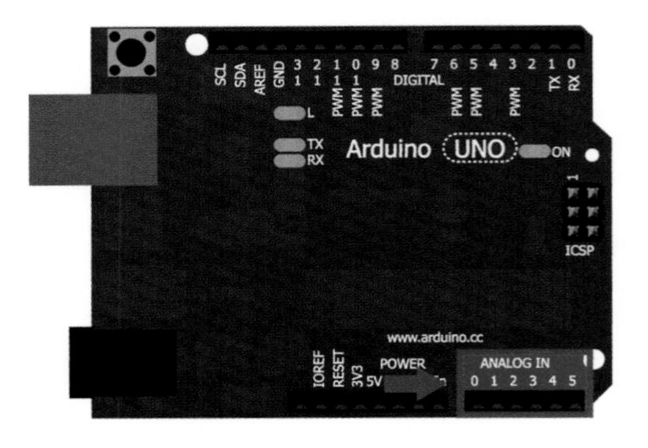

Figure 16-10 *The analog input pins of Arduino*

These pins can be used only as inputs, so you don't need to configure them in the setup block of your sketches. Try, for example, to connect a variable resistor to one of these inputs, as shown in Figure 16-11. Note that the center lead of the potentiometer and the wire that connects to it are both on the other side of the gap from all the other leads.

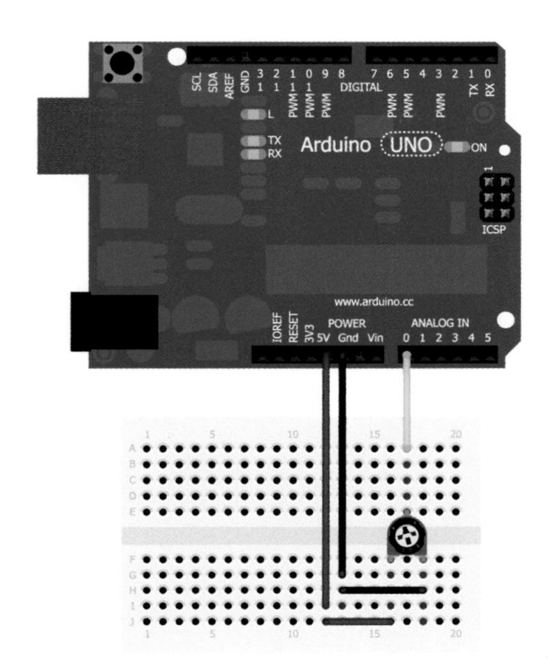

Figure 16-11 *A variable resistor connected to an analog pin*

A potentiometer or trimmer is a device whose resistance you change by rotating a knob, exactly like an early volume control (see Figure 16-12). To read the value of this and other analog inputs, you use the `analogRead` instruction, similar to the corresponding `digitalRead`, except `analogRead` is capable of returning a value between 0 and 1023. Of course, you can store this value in a variable:

```
int analogValue = analogRead(A0);
```

How can you use this kind of value? You could change the frequency that the LED blinks at (instead of a trimmer, you could use a proximity sensor and have the LED signal the approach of a potential danger with increasing intensity). You can connect an LED to pin 13 as before and write the code that corresponds to the behavior you are interested in.

We're cheating here by not including a resistor between the LED and one of the Arduino pins. Although this technically is stressful on the LED and the Arduino's pin, we've never broken an Arduino doing this. Earlier Arduino models had a built-in resistor on pin 13 (in newer models, that resistor is shunted off to the onboard LED), so we blame this on a habit we picked up in the old days and just haven't shaken off.

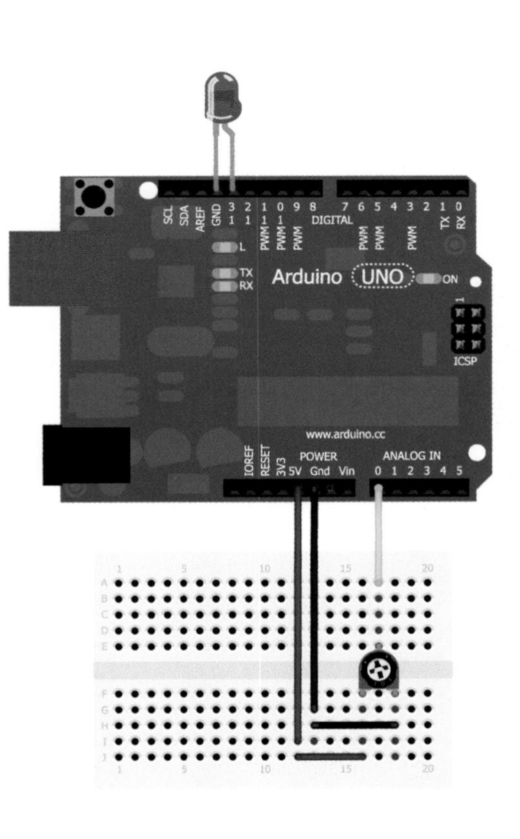

Figure 16-12 *An LED and a trimmer ready for use*

You already know how to do everything you need, so you just need to combine what you have learned so far. To show that there's an LED on pin 13, you'll write:

```
int led = 13;
void setup()
{
   pinMode(led, OUTPUT);
}
```

To change the frequency, you need to place the delay between one instruction and the next one stored in a variable, which takes the value read by the trimmer:

```
interval = analogRead(A0);
```

The complete listing of the sketch is shown in Example 16-7.

Example 16-7 *Using a trimmer to change the delay of a blinking LED*

```
int led = 13;
int interval = 1000;
void setup()
{
   pinMode(led, OUTPUT);
}
void loop()
{
   interval = analogRead(A0);
   digitalWrite(led, HIGH);
   delay(interval);
   digitalWrite(led, LOW);
   delay(interval);
}
```

Verify and upload the sketch, then try to vary the frequency at which the LED blinks by operating the trimmer or potentiometer knob. When you're using a variable resistor like this, it helps to set a threshold (i.e., a value below which you don't want anything to change), as shown in Example 16-8.

Example 16-8 *Ignoring all values below a certain threshold*

```
int led = 13;
int interval = 1000;
int threshold = 250;
void setup()
{
   pinMode(led, OUTPUT);
}

void loop()
{
   interval = analogRead(A0);
   if (interval <= threshold)
   {
      interval = threshold;
   }
   digitalWrite(led, HIGH);
```

```
    delay(interval);
    digitalWrite(led, LOW);
    delay(interval);
}
```

First, we define variables: `led` represents the pin to which the LED is connected, `interval` determines the frequency at which the LED will blink (initially equal to `1000`), and `threshold` is the value below which the system behavior doesn't change.

In the setup, the only thing you need to do is to set the LED pin as an output.

In the loop, you first read the value (`analogRead(A0)`): if it is less than or equal to the threshold (`interval <= threshold`), you set the interval variable to the threshold value (`interval = threshold`). If, on the contrary, the read value is above the threshold, you don't need to change anything (you just leave the interval at the value you read from `analogRead`), so there is no `else` block. Then you make the LED blink.

I want to shout it to the world!

The frequency at which the LED blinks gives us an indication of the value the sensor has read. How could you find out the exact value? One possible solution—and the simplest one—is to use the *serial monitor*, a tool built into the Arduino IDE that allows you to monitor activity on the Arduino's *serial port*. How is it used? It's like being on the phone with a friend: you first have to dial the number and call. In a similar way, to be able and use the serial port on Arduino you have to use the `Serial.begin` command in the sketch setup:

```
void setup()
{
   Serial.begin(9600);
}
```

The parameter `9600` indicates the speed at which you intend to communicate on the serial port, in this case 9,600 bits per second. To write to the serial port, use the `Serial.println` instruction, where you specify the desired text as a parameter. Basically, this is the equivalent of talking into the phone receiver (anachronistic, yet romantic). A simple sketch for writing the trimmer value to the serial port, keeping the same circuit as in the previous example, is shown in Example 16-9.

Example 16-9 *Printing the values read by a trimmer*

```
void setup()
{
   Serial.begin(9600);
}

void loop()
{
   Serial.println(analogRead(A0));
   delay(1000);
}
```

You have introduced a one-second delay to avoid having too many readings, which would make it harder to understand what is happening. This simple process of writing on the serial port allows Arduino to communicate with the rest of the world and will be very useful for future projects. As for now, we are happy to see the data value that was read. Verify and upload the sketch; then, to open the serial monitor, click the magnifier icon on the top right of the development environment, and select Serial Monitor from the Tools menu (Figure 16-13) or press the shortcut key Control-Shift-M (Windows, Linux) or Command-Shift-M (Mac).

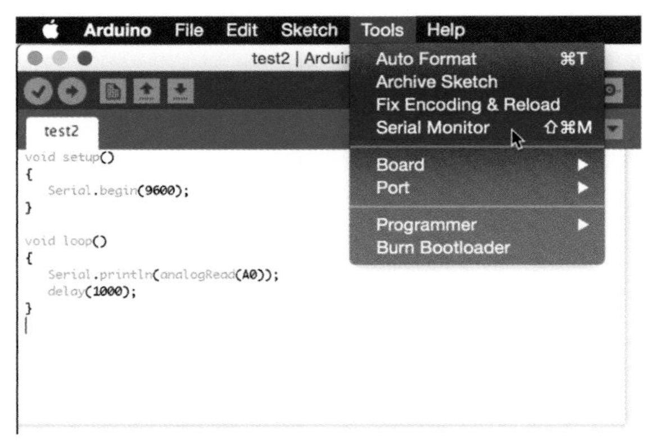

Figure 16-13 *Opening the serial monitor*

Once open, the serial monitor will display all the readings it receives from the Arduino (Figure 16-14).

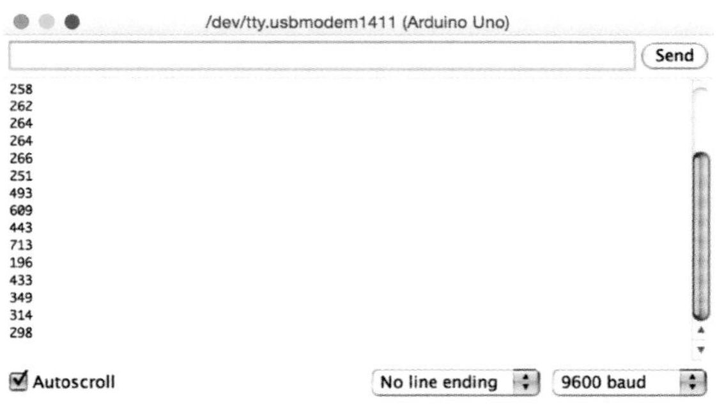

Figure 16-14 *Values on the serial monitor*

PWM is a bunch of pulses

Since you can read analog values with the `analogRead` instruction, you might be wondering if there's a corresponding `analogWrite` function and how you might use it. Yes, this function exists; you just need to find out on which pins we can use it. Most of Arduino's pins are digital only, and the few analog pins are input only. But if you look very carefully at Arduino's digital pins, you can see a few of them are marked with the letters *PWM*, which is an acronym for *pulse-width modulation*. (On some Arduino models, these pins are simply marked with a tilde ~, which roughly looks like the waveform of an analog signal.) These pins are firmly digital, yet under the right circumstances they can "pretend" to be analog.

These pins allow you to adjust the intensity of the outgoing signal: as if, rather than having a tap fully open or fully closed, you could decide how open it should be and adjust the water outflow. This is a very useful behavior if you need to operate particular devices like motors, for which you can adjust the rotation speed, or LEDs, for which you can adjust the brightness. How does this work?

It is basically the same concept behind traditional film projectors, which project 24 frames per second:

- If you project only black frames, the perceived image will be black.
- If you project only white frames, the perceived image will be white.
- If you alternate white and black frames, the perceived image will be grey. If you adjust the ratio, say, with two black frames for one white frame, the image will be dark grey. If the ratio is swapped, with two white frames for one black frame, the image will be a lighter grey.

The Arduino allows not just 24 shades, but 256, where at a value of 128, the output is right in the middle. Here's the syntax to use if we want to output 128 to pin 9:

```
analogWrite(9, 128);
```

The first parameter is the pin that concerns us, and the second one is the analog value we want to simulate. The `analogWrite(9, 0)` command corresponds to `digitalWrite(9, LOW)`, whereas `analogWrite(9, 255)` is equal to `digi` talWrite(9, HIGH). Figure 16-15 shows the representation of a PWM signal.

You can use this property to fade an LED up and out, by varying its intensity. First, you need to prepare a circuit where the LED is connected to a PWM pin, such as pin 9 (Figure 16-16).

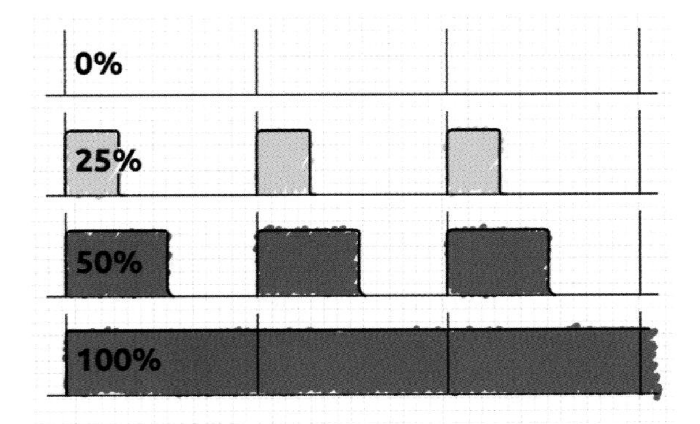

Figure 16-15 *Representation of different possible PWM cycles*

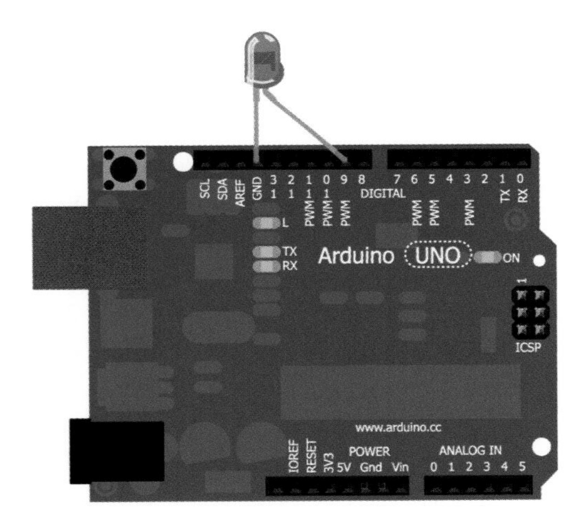

Figure 16-16 *An LED connected to a PWM pin*

That pin must be identified as an output in the sketch setup:

```
int led = 9;
void setup()
{
   pinMode(led, OUTPUT);
}
```

In the loop, you need to start with an intensity value of 0 and take it to the maximum (255). For this, you can use a cycle within the main loop block where the counter is also used to denote the LED intensity:

```
for (int i = 0; i <= 255; i++)
{
   analogWrite(led, i);
   delay(10);
}
```

To fade it out you set a similar process, decreasing the counter:

```
for (int i = 255; i >= 0; i--)
{
   analogWrite(led, i);
   delay(10);
}
```

The complete listing of this example is shown in Example 16-10.

Example 16-10 *An LED that gradually fades in, up, and out*

```
int led = 9;

void setup()
{
```

```
    pinMode(led, OUTPUT);
}

void loop()
{
    for (int i = 0; i <= 255; i++)
    {
        analogWrite(led, i);
        delay(10);
    }
    for (int i = 255; i >= 0; i--)
    {
    analogWrite(led, i);
    delay(10);
    }
}
```

If you wanted to, you could combine the two cycles, but the code would be a bit harder to understand. You can still try; have a look at it in the Fade example for inspiration (File→Examples→01.Basics→Fade).

If you need any help while writing an Arduino sketch, try the online reference. To view it, click Help on the menu, then Reference.

Some Exercises to Try

Let's close the chapter with a few things you should try on your own:

1. Make a circuit and a sketch where one LED blinks and another one switches on and off gradually, in the same cycle. When the first LED is on, the second slowly ramps up its brightness; when the second LED reaches the maximum brightness, the first LED gets turned off, and the second fades out again. Repeat.
2. Make a circuit and a sketch in which pressing the button toggles between two modes: one in which the LED fades, the other in which the LED blinks.
3. Make a circuit and a sketch where you can fade an LED in and out through a trimmer or potentiometer, using a maximum of 10 possible brightness values. Note: bear in mind the range of values of input (0–1023) and output (0–255) analog signals! A little hint: the `map` function, which you can find in the online reference, might help.
4. Add a trimmer to the circuit in the previous exercise and edit the sketch so that the delay varies depending on the trimmer value. Plan a minimum delay of 50 milliseconds and a maximum of 2.52 seconds.

17 Expanding Arduino

Now that you understand the basics of Arduino, you can have fun and use it for many different purposes. Many ideas can be found online, and some even come with full documentation on how to reproduce them. In Chapter 16, you saw that an Arduino is capable of interacting with sensors and actuators; let's have a more in-depth look at these components.

Reading the World: Sensors

A sensor is a component that allows you to measure a physical quantity or detect an event. The simple button you used in earlier exercises is also a simple sensor, which detects whether someone is pushing it. There are many different kinds of sensors. How do they work?

Thermistors

Thermistors (Figure 17-1) are based on the premise that some materials vary their electrical resistance as their temperature changes. The change is consistent, and can therefore be easily quantified: a resistance of *X* always means a temperature of *Y*. The ratio can be direct (the resistance increases as the temperature increases) or inverse (the resistance decreases as the temperature increases).

Figure 17-1 *A thermistor*

Therefore, reading temperatures in Arduino is easy. Since a thermistor is an analog sensor, you'll connect it to one of the ports between A0 and A5, then add a resistor and create a voltage divider like the one pictured in Figure 17-2. In this way, part of the supply voltage (5V) will be applied to the thermistor, and part of it to the resistor. If you read the incoming voltage value on the analog pin (i.e., the voltage at the ends of the thermistor) and you know the relationship between resistance and temperature, you can determine the temperature.

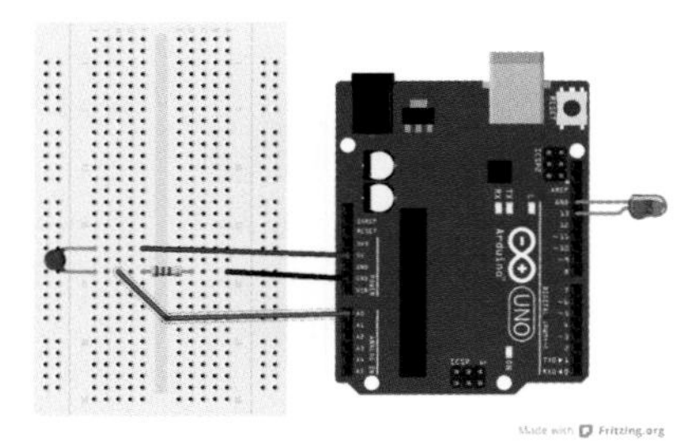

Figure 17-2 *A thermistor in action*

In the sensor documentation you can usually find all the information you need to convert the readings of Arduino inputs into the quantity you want.

Photoresistors

Other materials react to light. Their resistance changes when they are illuminated. These materials are used to make photoresistors (Figure 17-3): small disks that change their resistance as the light level changes.

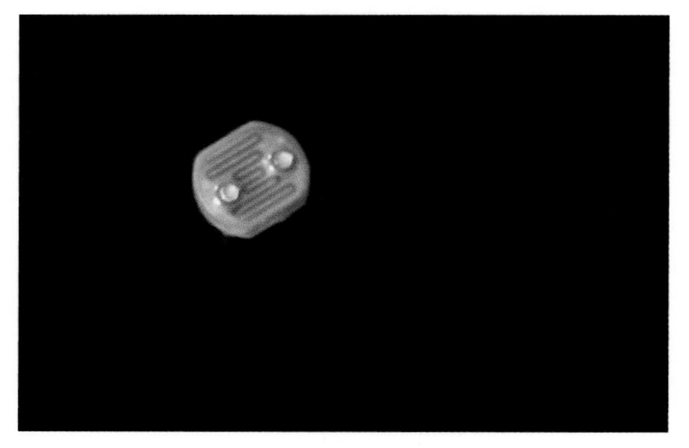

Figure 17-3 *A photoresistor*

The circuit needed for a photoresistor (Figure 17-4) is identical to the one you have just seen for the thermistor. Many other kinds of analog components are connected in the same way. You typically want to use a resistor that matches the range of the photoresistor in your voltage divider. For a photoresistor with a maximum resistance of 10K, use a 10K resistor.

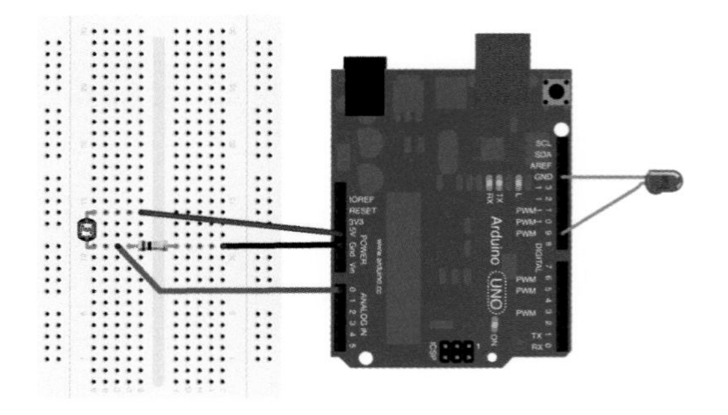

Figure 17-4 *A photoresistor is connected like a thermistor*

With both a photoresistor and thermistor, you can use `analogRead`, as described in "Beyond Digital" on page 152.

Other Kinds of Sensors

There are other materials that change their resistance if put under pressure, bent, or exposed to humidity or water: the same principle is valid for all of them, and each one can be used to make a sensor. There are also more complex sensors (Figure 17-5) that are able to measure the presence of specific gases and substances, the earth's magnetic field, the acceleration sustained by an object, or other phenomena. Basically, there is probably a sensor for anything you want to measure. Even radios and cameras can serve this purpose. For instance, with a webcam and a series of software libraries for image editing you can identify colors, shapes, and even people (though you'd need a Raspberry Pi or personal computer to have the computing power needed to process such things).

Figure 17-5 *Some kinds of sensors*

Other sensors can be queried through a serial port, and there are libraries that allow you to use them with Arduino. There are sensors of all sizes and different prices, hence the only limit is your imagination. Here are some exercises you could try with sensors:

- Make a circuit that switches a series of LEDs on when the ambient light falls below a certain threshold (for example, when you struggle trying to read a book).
- Modify that circuit so that the LEDs fade up gently as the ambient light fades out.
- Using your favorite technology (such as a laser cutter), make a table lamp that uses that circuit.

Changing the World: Actuators

Some components carry out a complementary function to sensors, in that they act on their surrounding environment, instead of merely reading it. These components are called *actuators*, and they usually turn electricity into movement, light, heat, or sound. You have already seen an example of an actuator when you made your Hello World of electronics; that is, when you switched an LED on, which is one of the simplest actuators: when crossed by current it emits photons, turning electricity into light.

Buzzers

Another kind of actuator is the *piezoelectric transducer*, more familiarly called a *buzzer*. The diaphragm of the device expands and contracts as the voltage changes, thus producing a vibration. Its frequency depends on the variation of the applied voltage; basically, it turns electricity into vibrational mechanical energy. You can try to use one to play a short tune with Arduino if you connect the components as shown in Figure 17-6.

To make your buzzer buzz, you will use the `tone` function, which accepts `pin`, `frequency`, and `duration` as parameters. The last parameter is optional, but if not specified the buzzer will keep on buzzing until you use the `noTone` function. Example 17-1 shows an example sketch.

Example 17-1 *You can make Arduino play a tune*

```
void setup()
{
   pinMode(9, OUTPUT);
}
void loop()
{
   tone(9, 131, 100);
   delay(500);
   tone(9, 147, 100);
   delay(500);
   tone(9, 165, 100);
   delay(500);
   tone(9, 175, 100);
   delay(500);
   tone(9, 196, 100);
   delay(500);
}
```

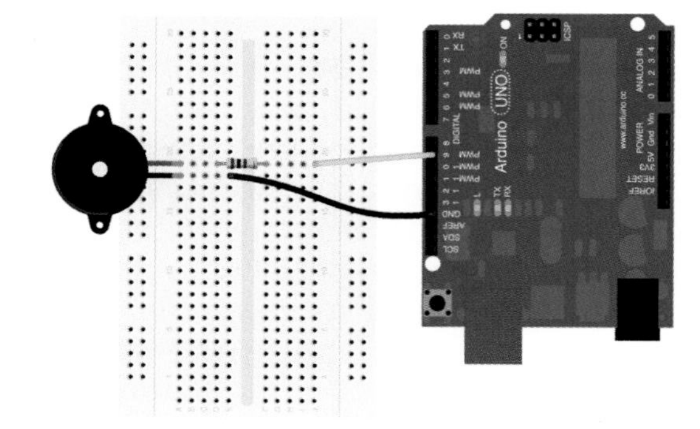

Figure 17-6 *Connecting a buzzer to Arduino with a 100 ohm resistor*

Warning

On boards other than the Arduino Mega, the `tone` function interferes with the PWM on pins 3 and 11.

Servos

Another example of an actuator that turns electricity into mechanical energy is the *servo*, a particular kind of motor capable of moving an arm connected to it into a specific position and holding it there. Usually, it can perform rotations of up to 90 or 180 degrees. You could use it for many purposes, such as turning a doll into an *animatronic* machine with independent movement capability. Let's try to make a circuit with a servo and a potentiometer, which you will use to control it (Figure 17-7).

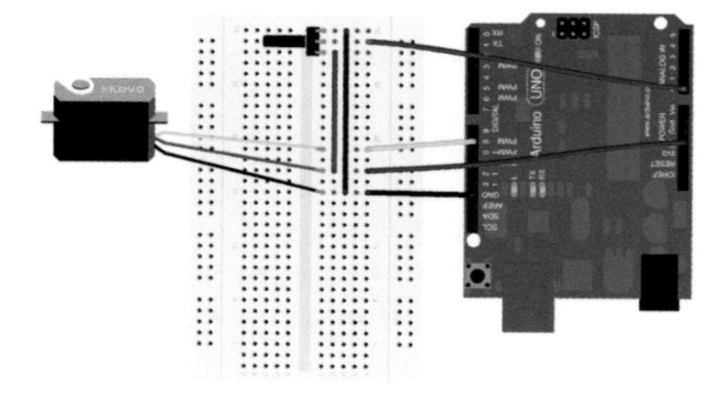

Figure 17-7 *A servo controlled by a potentiometer*

Thanks to the Servo library, which manages the PWM signal that drives the movable arm, using a servo is very simple (Example 17-2).

Example 17-2 *Controlling a servo with a potentiometer*

```
#include <Servo.h>
Servo myservo;
int position = 0;
void setup()
{
   myservo.attach(9);
}

void loop()
{
   int pos = analogRead(A0);
   position = map(pos, 0, 1023, 0, 180);
   myservo.write(position);
   delay(100);
}
```

After you include the library with the `#include<Servo.h>;` statement, the sketch creates a variable `myservo`, which represents your servo. In `setup`, the sketch attaches `myservo` to pin 9.

In `loop`, the sketch takes the value read by the sensor connected to the potentiometer and converts it into degrees, so that the maximum value, 1023, corresponds to 180 degrees. The sketch then tells the servo to rotate the movable arm to the determined position, and wait 100 milliseconds to allow for the rotation. Now you can modify all the dolls you have at home! You can even cut out a board with the names of your colleagues and use the `random` function to generate a random number and move the servo arm, so that it indicates who's due to bring in donuts the next day:

```
int num = 10; // # of colleagues
int chosen = random(num);
position = map(chosen, 0, num, 0, 180);
myservo.write(position);
```

Before you move on, try some exercises:

- Complete the donuts sketch and change it so that you'll have to bring them in less often (after all, you are the one who creates the hack, right?).

- Change that sketch so that the arm stops at the center of each section, in order to avoid arguments when it is not clear who's going to bring in the donuts the next day.

Strong Currents

The Arduino can manage current up to 40 milliamps per pin: this value is very small if you consider that a simple MP3 player requires something like 100 milliamps, and an electric oven around 30 amps. If you wanted to use an Arduino to command the oven to start cooking dinner, there's no way the Arduino could supply the power the oven needs. And any attempt to push 30 amps through an Arduino would end up with a roasted circuit board, not a cooked dinner.

This does not mean that you *can't* connect your oven, or a lamp, or even a 12 volt motor to an Arduino, only that you have to go about it a different way. You simply have to use specialized electronic components to separate the Arduino's weak current from the much stronger current in the rest of the circuit.

For this purpose you can use a device called a *MOSFET* (an acronym for *metal-oxide-semiconductor field-effect transistor*) whose functioning is very simple. Think again of our comparison with water pipes (Figure 17-8): imagine an irrigation valve that is activated when the timer linked to it sends the pertinent signal. The valve has three ports: a *source* flowing in, a *drain* flowing out, and a *gate* that controls whether the valve is open or closed.

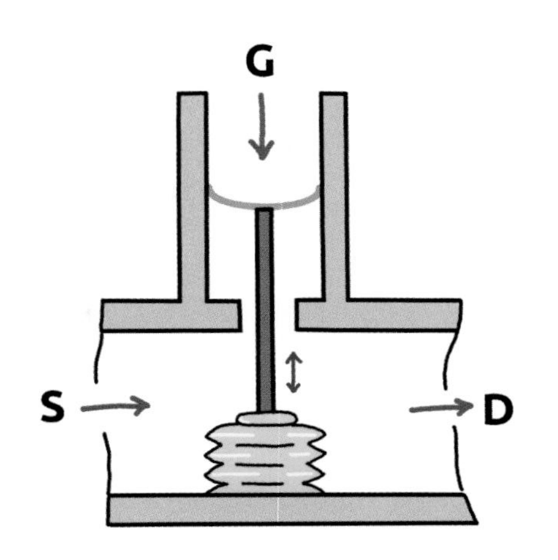

Figure 17-8 *The hydraulic analogy of a MOSFET*

The MOSFET works in an analogous way (Figure 17-9): when a (weak) signal gets to the gate, the MOSFET allows the (strong) current to flow from the source to the drain.

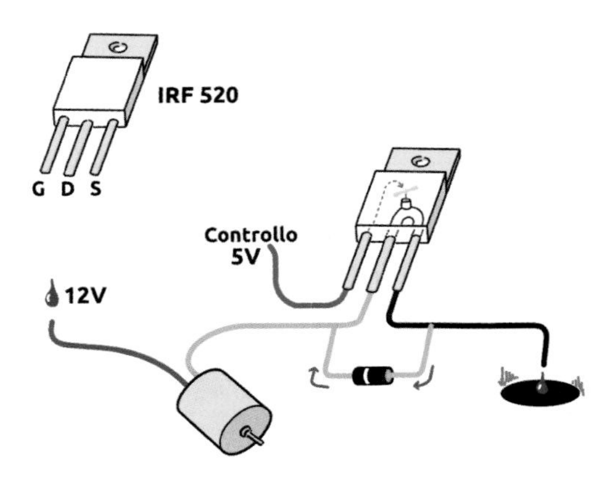

Figure 17-9 *Functioning scheme of a MOSFET with a protection diode for inverted currents (indicated in red)*

With a MOSFET you can easily control actuators that need more current than can be managed by an Arduino alone—for example, an electric motor or a stepper motor—with the additional advantage that you can use voltages different from the ones supplied by the board. With a properly rated MOSFET (one that could handle the voltages and amps required) it should be relatively easy to control a large device from a small Arduino. Figure 17-10 shows the circuit for this. Note the use of the diode to protect Arduino from any current that the motor produces itself (since a motor can also be a generator when you spin it by hand).

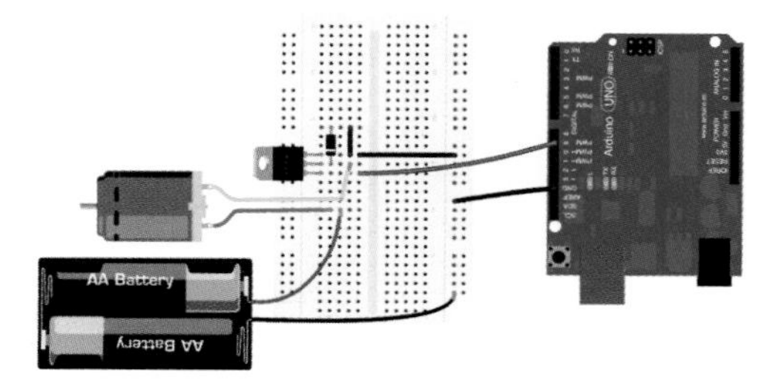

Figure 17-10 *Arduino controls a direct current motor powered at 3V*

Shields

Arduino is not only very versatile, but expandable too. To use particular sensors or actuators, it is often necessary to create quite complex circuits comprising microprocessors and specific components. To meet the beginners' needs and also spare experts some time, many producers have created a series of boards called *shields*, only a few of which are shown in Figure 17-11, that recreate these circuits and plug directly into the Arduino. Most of these shields are designed to be modular, so you can integrate Arduino with even more shields and create a stack that resembles a huge sandwich and has much more functionality. For example, you can create a *green station* to monitor environmental parameters (using one dedicated shield with specific sensors), store the data (using another shield with external clock and data logger) and publish them (using a Wi-Fi or Ethernet shield).

Figure 17-11 *Shields for Arduino*

Each shield has its own specific functions and its own documentation; some also require the use of additional software libraries to access these new functionalities from your development environment. You can buy shields in most places where Arduinos are sold.

Smart Textiles

Technology allows you to use new tools and materials to give life to your creations. Those that you can wear and that allow you to combine electronics with fashion are known as *wearables*. In fact, some types of materials, in particular conductive textiles (Figure 17-12) and conductive threads, enable you to create real electrical circuits; these fabrics are generally known as *smart textiles*, even if the most correct term would be *e-textiles*.

A further advantage offered by these textiles is that they can be washed without being damaged—although, before starting the washing machine, you must not forget to remove the components, which may otherwise get damaged.

When you're developing wearables, our suggestion is to carry out the project iteratively, testing that everything is being processed in the proper way as you proceed toward the final result.

Figure 17-12 *A conductive textile can replace the electrical threads in a circuit*

One of the problems you face with wearables concerns the power source—most of us would find being constantly attached to a power adapter rather confining, and constantly having to replace batteries annoying and wasteful. Another limitation is that, although Arduino is a compact board, if you have to attach it to a jacket or pair of pants it may feel bulky and awkward. Here the solution is *LilyPad Arduino* (Figure 17-13), a simplified version of Arduino specifically designed to be used with wearables. The LilyPad has some eyelets covered with a conductive material that you can use to stitch your board to fabrics with some conductive threads, creating a real circuit. You can also use a small rechargeable battery with the LilyPad.

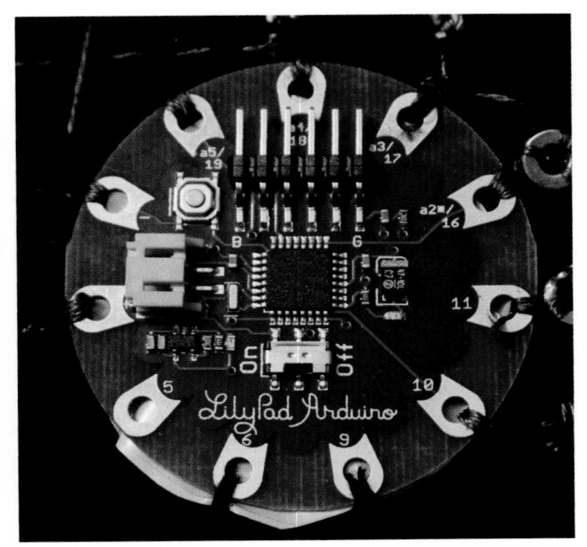

Figure 17-13 *A LilyPad circuit attached to a fabric with conductive thread*

One big difference between the LilyPad and Arduino Uno is that the LilyPad has fewer pins than the standard Arduino, and these pins are well separated from one another in order to avoid any accidental short circuit. Another difference is that there is no classical USB connector, so it is necessary to use a specific *programmer*—that is, a small board that converts signals from USB to serial communication. In addition, to solve the problem of power consumption, the LilyPad is equipped with a button that can switch off the circuit when it's not in use; however, when you connect the LilyPad to a computer, the LilyPad computer will always be powered regardless of the position of the switch.

There are also LEDs designed for wearables (Figure 17-14). They are smaller and are assembled on small boards with two holes for stitching them, just like LilyPads. Of course nothing prevents you from using traditional LEDs, and the possibility of choosing between different solutions for a particular project offers you more freedom of use.

Figure 17-14 *Sewable LEDs*

Before you use a LilyPad, it is advisable to test the circuit with a classical Arduino because it is much easier to add, move, or remove links. Once you have tested everything, you can load the sketch on your LilyPad, connect it to the rest of the circuit, and test your fantastic artifact!

Raspberry Pi

18

Sometimes, an Arduino is not enough. There are projects that require flexibility and calculating power that the Arduino's small chip can't provide. If you wanted to, say, re-create an 80s-style game console, mine for whichever cryptocurrency is in fashion at the moment, or drive a high-definition video display, the Arduino can't really hack it. For cases like these, you'd turn instead to the Raspberry Pi.

Raspberry Pi is a small, inexpensive, full-fledged computer on a single circuit board, complete with operating system, keyboard and mouse support, and a graphical user interface (GUI) with windows and menus. With the quad-core processor powering the latest generation of Raspberry Pi, it is similar to a smartphone or tablet in terms of computational power. It can be connected to a monitor, a mouse, a keyboard—and many USB devices. Plus, it can connect to the Internet, and execute programs written in languages such as C, Java, and Python.

By this measure, Arduino and Raspberry Pi are not really comparable. They both have a series of pins that can be connected to external circuits, yet each specializes in carrying out different tasks. Arduino is, essentially, just a microprocessor with some additional hardware for handling various types of input and output. All it can do is to run the same small program over and over until the power gives out (or you put in a new program). On the other hand, because the Raspberry Pi is a full computer, it is not as reliable as Arduino in terms of interacting with the hardware in real time.

Component Check!

As soon as you open your Raspberry Pi box, you see a small printed circuit board, a little bigger than Arduino.

Let's take a closer look at Raspberry Pi (Figure 18-1):

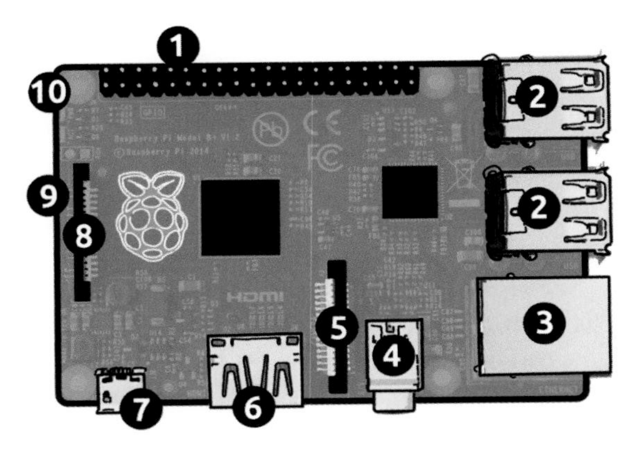

Figure 18-1 *The most important parts of a Raspberry Pi*

1: GPIO (General-Purpose Input/Output) header
These pins can be used as digital inputs or outputs, similar to the pins on Arduino. The single-core Raspberry Pi Models A+ and B+ as well as the quad-core Raspberry Pi 2 Model B have 40 pins, in two 20-pin rows. Unlike the Arduino, the Raspberry Pi has only digital pins, with no analog inputs. Its pins operate at a voltage of 3.3V, as opposed to Arduino's 5V; keep that in mind when you connect hardware to the GPIO pins. Too much voltage can damage the pins or the entire Raspberry Pi.

2: USB ports
The Model B+ and Raspberry Pi 2 have four USB 2.0 ports. Just about any *class* of USB device (e.g., mouse, keyboard, Wi-Fi, Bluetooth, and GPS) will work with the Raspberry Pi. When things don't work, the chief culprit is probably power. With your Pi plugged into a strong enough (2 amps or more) USB power supply, the Raspberry Pi ports can provide a maximum of 500 mA to USB devices. If a USB device requires more than 500 mA, or if you're powering your Pi with less than 2 amps, that USB device may not work reliably. For this reason, it is better to connect a USB device to a separately powered USB hub, in order to isolate the Raspberry Pi from the most demanding gadgets.

3: Ethernet connection
The Ethernet connector is present only on B type (B+, Raspberry Pi 2) boards. In today's wireless world, it could be hard to find a wired network jack to connect Raspberry Pi to, so it might be much easier to connect a compatible USB Wi-Fi dongle (*http://elinux.org/RPi_USB_Wi-Fi_Adapters*). To connect to a Wi-Fi network after you've connected a Wi-Fi module, start the GUI (see "The Graphical User Interface" on page 175) and click Menu→Preferences→WiFi Configuration, then click the Scan button to find a network to connect to.

4: Analog audio output
Besides the digital audio found on the HDMI output, there is a lower-quality analog audio source signal available on this 3.5 mm audio jack connector. You could also connect headphones, but you can hear the signal better with a pair of amplified speakers.

5: CSI (Camera Serial Interface)
This is the connector for the Raspberry Pi camera accessory. There are two versions of the camera: the normal camera model and an infrared version (the Pi Noir Night Vision Camera Module).

6: HDMI output
With a simple HDMI cable that costs only a few dollars, you can use this connector to plug the Raspberry Pi into a monitor or a TV.

7: Power in
This is the micro USB port used to supply power to the Raspberry Pi. It is possible to use a smartphone charger to power the Pi, but be sure it puts out 5V and at least 700 mA. Make sure you use enough power. An underpowered Raspberry Pi tends to reset itself in the middle of the most critical operations, when the most electrical power is needed.

8: DSI (Display Serial Interface)
With this connector you can (eventually, when the official display panel is released) connect a display via a flat, flexible cable similar to the cable used to connect the camera.

9: Micro SD slot
On the underside of the Raspberry Pi, there is a housing for a Micro SD card, which is what the Raspberry Pi uses in place of a hard disk. Just as with the hard drive on a larger computer, the SD card contains the operating system for the Raspberry Pi. You must use a good quality card (at least class 4). SD cards higher than class 8 are much faster, but some users have complained

that they're not as stable; try to stick to class 4 or class 6 cards.

10: Status LEDs

There are two status LEDs here, PWR (power) and ACT (activity). When the power LED is on and stable, Raspberry Pi is being properly supplied with power. The activity LED shows when the Micro SD card is being accessed.

Getting Started

Let's get started with the Raspberry Pi. Connect a mouse and a keyboard to the USB ports. For the video, you can use an HDMI digital input TV or monitor and an HDMI cable.

As we've said, in place of a hard disk, the Raspberry Pi uses a Micro SD card on which you need to load the operating system. You'll need to obtain a Micro SD card (8GB or larger is suggested, but it will work with a 4GB card). Grab the NOOBS installer from the Raspberry Pi downloads page (*http://www.raspberrypi.org/downloads/*) and unzip it onto your Micro SD card. Check out the NOOBS setup guide (*http://www.raspberrypi.org/help/noobs-setup/*) to help you with the installation.

To switch the Raspberry Pi on, connect a power supply. When it starts up, the NOOBS installer will ask you which operating system to install. Click the checkbox next to Raspbian and then click the Install icon. After the installation is complete, the Raspberry Pi will restart, and you'll see a black screen scrolling with fast-moving white text, until the raspi-config configuration program screen appears, as shown in Figure 18-2.

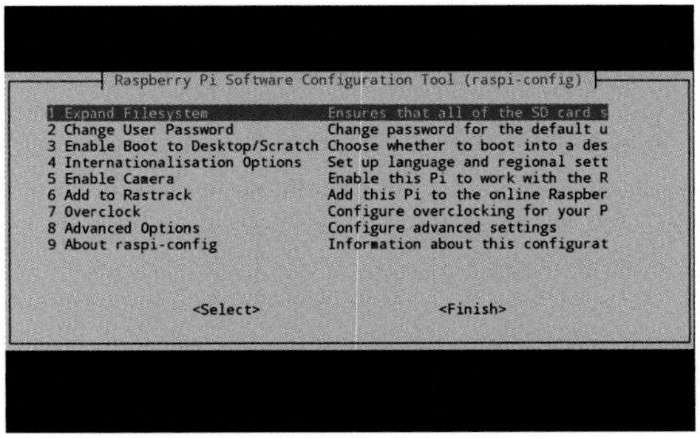

```
Raspberry Pi Software Configuration Tool (raspi-config)

1 Expand Filesystem                Ensures that all of the SD card s
2 Change User Password             Change password for the default u
3 Enable Boot to Desktop/Scratch   Choose whether to boot into a des
4 Internationalisation Options     Set up language and regional sett
5 Enable Camera                    Enable this Pi to work with the R
6 Add to Rastrack                  Add this Pi to the online Raspber
7 Overclock                        Configure overclocking for your P
8 Advanced Options                 Configure advanced settings
9 About raspi-config               Information about this configurat

          <Select>                          <Finish>
```

Figure 18-2 *When you first start the system, the raspi-config program will help you configure your Raspberry Pi*

The configuration operations are rather simple. Choose Internationalisation Options with the arrow keys, and then press Enter or Return. You should set your Locale to one appropriate for where you are (for example, in the US, you probably want "en_US.UTF-8 UTF-8"). Return to the Internationalisation Options and set your time zone and keyboard layout.

The keyboard layout selection may involve a few steps. US users will probably need to choose either Generic 101-key PC or Generic 104-key PC, and if English (US) is not available, choose other, then choose English (US).

When you're done configuring your Pi, press Tab until Finish is highlighted, then choose the option that reboots your Pi.

When the Raspberry Pi boots up next, it will start up to a black screen with a prompt labeled `login:`. At the prompt, authenticate yourself by typing in `pi` for the username and `raspberry` for the password. You'll be greeted by a *shell prompt* that indicates the shell (an interactive tool for running programs and sending commands to the Pi) is ready for you to type commands:

```
pi@raspberrypi ~ $
```

The prompt is made up of several parts:

`pi`
: This is your username.

`@raspberrypi`
: This indicates that your Raspberry Pi's hostname is raspberrypi. Read together with the preceding part of the prompt, it indicates that you're logged in as user "pi," at ("@") the host named "raspberrypi."

`~`
: This shows your current working directory. At the moment, you're in the home directory (*/home/pi*), and ~ is shorthand for your home directory.

`$`
: This is the shell prompt. When you see it, it means the shell is ready for you to send it commands.

But...why are you operating in a black-and-white text environment? Doesn't the Raspberry Pi have a graphic environment? Yes, but it does not start automatically. Before you launch it, it is worth having a look around the text environment and getting familiar with the use of the display and some of the Linux commands.

One of the fundamental metaphors in the computer world is the concept of *files*, and *folders* in which to organize those files. The folder metaphor, illustrated in Figure 18-3, imagines that your computer's hard drive (or SD card, in Raspberry Pi's case) is an old metallic filing cabinet with many drawers full of folders, containing many individual documents (files) or even smaller folders (subfolders).

Another metaphor is the tree structure, used to define the hierarchy under which data are stored (Figure 18-4). To make your way through this hierarchy, you have to type commands, which are interpreted by the shell (the program that starts running right after you log in).

The Right Way to Turn Off

The Raspberry Pi does not have an on/off power switch or a reset button. As soon as you supply power to the board, the operating system loads. If you remove power from the board (by literally pulling the plug), the board switches off.

But it is never a good idea to switch off a computer this way, while the operating system (OS) is working. Nearly every OS performs operations in the background, such as writing data to the SD card, which we as users are not aware of. If you remove power suddenly, the OS might be caught in the middle of one of those background tasks, with the result that the SD card might become corrupted, and unbootable next time. The best way to switch Raspberry Pi off is to launch a specific command (`sudo shutdown` or `sudo halt`) from the shell.

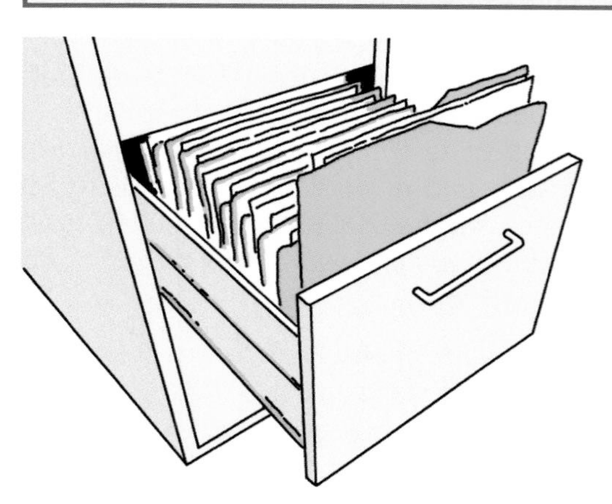

Figure 18-3 *A physical archive with files and folders*

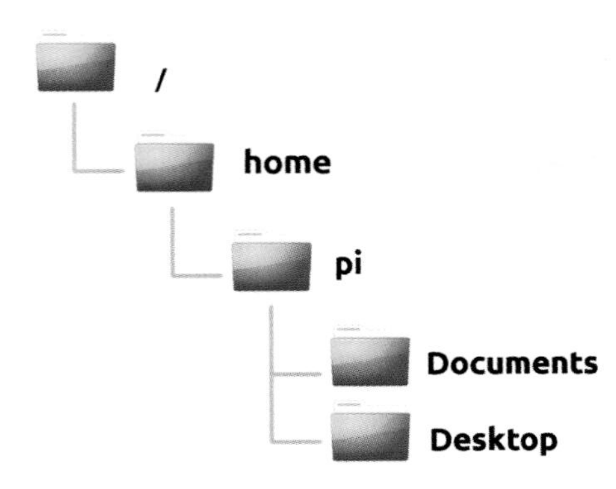

Figure 18-4 *An example of the folder hierarchy of Raspberry Pi*

Another term for folder is *directory*, which is more commonly used on Linux systems (Linux is the operating system that runs on the Raspberry Pi). The main directory, which contains all the others and is at the top of the hierarchy, is called *root* and is represented by the symbol / (*slash*).

Basic Shell Commands

To find out which branch (directory) of the tree you are in, type `pwd` followed by Enter or Return. `pwd` stands for *print (name of) working directory*, and displays the path from the root to the current directory:

```
pi@raspberrypi ~ $ pwd
```

When you see an example like this one, don't type the $ or the text leading up to it. The $ is the shell prompt, and you need to type your commands after the prompt.

The system will reply:

```
/home/pi
```

How do you know what is contained in the directory you are in? You can list the current files by using the command `ls`, which stands for *list*:

```
pi@raspberrypi ~ $ ls
```

On a freshly installed Raspberry Pi, the system will reply with the following (if you've created files or directories on your own, you'll see something different):

```
Desktop  python_games
```

You can get more information about the files by using the parameter `-l`, which stands for *long*:

```
pi@raspberrypi ~ $ ls -l
```

The system will respond with more details about the two directories:

```
total 8
drwxr-xr-x 2 pi pi 4096 Jan 31 16:26 Desktop
drwxrwxr-x 2 pi pi 4096 Dec 31  1969
python_games
```

To move your focus from one directory to another, you can use the `cd` command (which stands for *change directory*) followed by the name of the directory you want to access:

```
pi@raspberrypi ~ $ cd Documents
```

This tells you to move to the *Documents* directory. If you were to then run an `ls` command, you'd see the files in the *Documents* directory.

Directories can be nested. Inside the *Documents* directory might be a *Projects subdirectory*, for your project-related documents. To go there from the *Documents* directory, you can tell the OS `cd Projects`.

On a freshly set up Raspberry Pi, there won't be a Documents directory. You could create it (and the Projects subdirectory) with this command: `mkdir -p ~/Documents/Projects`*. You'll see more about the* `mkdir` *command in* *"Operations on Files and Directories" on page 172.*

To back out of a nested subdirectory into the directory that contains it, you can use a shortcut. The higher-level directory is represented by `..` (two dots), while the current directory is represented by `.` (one dot).

So, to go one level back you'll use `cd ..`, like so:

```
pi@raspberrypi ~/Documents $ cd ..
```

If you are in the directory */home/pi* and you use the `cd ..` command, you will go to the */home* directory.

The shell remembers the last commands you have given it: if, at the prompt, you press the up arrow or down arrow keys, you can run through the list of commands you have used in the order you have input them. This is very useful when you need to repeat a command several times. Another helpful feature of the shell is the *autocomplete* function for the commands: you just need to type in the first letters of a command, then press the Tab key, and the shell will complete it for you, saving you from typing the whole name. Of course, if there are multiple commands that start with the same letter, you might have to type enough of the command so that autocomplete knows which one you mean. You can also press Tab twice to see all the possible commands.

Operations on Files and Directories

Let's start by creating a directory with the `mkdir` (*make directory*) command, followed by the name you want to give to the directory. We're showing two commands here; the first one is to make sure you change back to your home directory before you try the rest of the exercise (if you type `cd` all by itself without any arguments, it changes to your home directory):

```
pi@raspberrypi ~ $ cd
pi@raspberrypi ~ $ mkdir test
```

Now, move into the directory and create an empty file with the `touch` command; then, with `ls`, check whether the system has actually created the file:

```
pi@raspberrypi~ $ cd test
pi@raspberrypi~/test $ touch hello.txt
pi@raspberrypi~/test $ ls
hello.txt
pi@raspberrypi~/test $ ls -l
total 0
-rw-r--r-- 1 pi pi 0 Feb 15 19:42 hello.txt
```

To copy a file, use the `cp` command, for *copy*, followed by the name of the file you want to copy and the destination you want to copy it to. Copy the *hello.txt* file to the *home/pi* directory:

```
pi@raspberrypi ~/test $ cp hello.txt /
home/pi
```

Because the home directory is one level higher than the current one, you may also write:

```
pi@raspberrypi ~/test $ cp hello.txt ..
```

And because ~ can be used as a shortcut for your home directory, you could also write:

```
pi@raspberrypi ~/test $ cp hello.txt ~
```

The `mv` command, which means *move*, is used to rename or move files. To rename your *hello.txt* file to *pi.txt*, you'll type:

```
pi@raspberrypi ~/test $ mv hello.txt pi.txt
pi@raspberrypi ~/test $ ls
pi.txt
```

To move it to the higher-level directory:

```
pi@raspberrypi ~/test $ mv pi.txt ..
pi@raspberrypi ~/test $ ls
pi@raspberrypi ~/test $ ls ..
```

You can delete the file with the `rm` command, which means *remove*. When you're working from a shell, there is no trash directory where deleted files are moved to; moreover, Linux is a system that obeys and does everything it is asked, including completely deleting all data and the operating system itself, so you'd better pay great attention to what you do:

```
pi@raspberrypi ~/test $ rm ../pi.txt
```

From your shell, you can even modify a file. One of the simplest editors you have at your disposal in Linux is nano, shown in Figure 18-5.

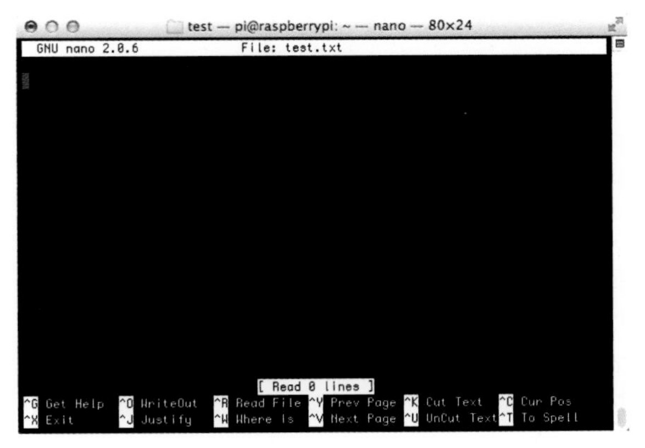

Figure 18-5 *The main screen of the nano editor*

In order to edit a file with nano, use the command nano, followed by the filename:

```
pi@raspberrypi ~/test $ nano pi.txt
```

One great thing about nano is that the file doesn't already have to exist; nano will create it if necessary. Type the phrase `Hello World!` into the file. After editing the file, type Control-O followed by Enter or Return to save the file. Type Control-X to exit the editor.

On Linux, it is possible to read a file's content without using an editor; the command `cat` (for *concatenate*) prints the file contents to the console:

```
pi@raspberrypi ~/test $ cat pi.txt
Hello World!
```

If the file is very long, you will see only the last part that fits in the terminal window. To make reading easier, you can use a text viewer called `less`:

```
pi@raspberrypi ~ $ less pi.txt
```

With `less`, you can scan through lines and pages of the file, or jump to the beginning or to the end of the text. To move within the text, use the up and down arrow keys, space, and Enter (or Return). To exit `less`, press the q key. When you press h, a short manual of the available commands is displayed.

The Raspberry Pi is always executing a number of programs in the background; to find out what they are use the `ps` (*process status*) command. If the command is followed by some parameters, you can get more or less information about the ongoing processes; the `-ef` option will display a detailed list of processes in the whole system.

```
pi@raspberrypi ~ $ ps -ef
```

Generally, if you're looking for any kind of information about the operating system, two programs are here to assist you: `apropos` and `man` (for *manual*). `apropos` will search for a string in the title of all manual pages, and `man` will show the given manual. Since most manuals are larger than one screen, `man` automatically uses `less`.

The command `man apropos` will display the manual for how to use `apropos`. By using those programs with other commands, you will find lots of information about all aspects of the installed software.

Redirection

On Unix-based systems, like Linux, each program is designed as a "black box" that receives input from a source (the default source is the keyboard), and produces output to a destination (the default destination is the shell output/computer screen, referred to as the *standard output*). These commands can be interconnected to create command chains that can perform even complex operations. To connect two commands, for example `ls` and `less`, you need to use a *pipe*—that is, the | symbol. If a directory contains many files, the `ls -l` output will take up many screens, but if you move it through `less`, you'll be able to easily read it:

```
pi@raspberrypi ~ $ ls -l | less
```

Instead of using the standard output, you can direct the output of a program elsewhere; for example, to write the output to a file, use the symbol > (greater than):

```
pi@raspberrypi ~ $ ls -l > list.txt
```

You won't see anything appear on screen, which is fine, because you told the operating system to output `ls -l` to the file *list.txt*. You can take your time and read the file later. Each time you repeat this command, the operating system overwrites the *list.txt* file. If you want to add the latest output to what is already in the file without overwriting it, you need to use the >> (double greater than) symbol:

```
pi@raspberrypi ~ $ ls -l >> list.txt
```

The command inputs work in a similar way. To redirect an input, use the < (less than) sign. This causes the command to read from the file rather than the standard input (which more or less corresponds to your keyboard):

```
pi@raspberrypi ~ $ cat < list.txt
```

In this case, the `cat` command operates on the information found in the *list.txt* file, just as if you had typed every character manually.

Most Linux commands that can operate on a file with the < redirection operator can also work just fine without it, if you supply the filename as an argument, i.e., `cat list.txt`*. You'll find that in Linux, there's often more than one way to do something.*

The World of the Superuser

In Linux, every individual who uses the system can be given her own user account, such as the *pi* account you've been using all along. But there is also a "superuser" called *root* built into the system, capable of performing any operation without limits: root can delete protected files and directories, even the entire disk. Using that much power on a Raspberry Pi is not recommended, so you would normally operate as a common user, with limited power. This limits the damage if you typed a wrong command.

But sometimes you need to perform operations as the root user. To make your life simpler, the sudo system (*superuser do*) has been created: with this, a "regular" user can launch superuser commands, such as switching the system off. To perform superuser commands, you need to place the word `sudo` before the command. Depending on the specific setting of the operating system, using `sudo` might require your password before the command is executed. By default, it does not ask for a password on the Raspberry Pi.

The following command will shut down your Raspberry Pi:

```
pi@raspberrypi ~ $ sudo halt
```

Of course, the `sudo` trick is not valid all the time and for everyone: the superuser needs to first include you within the list of people allowed to use `sudo` (the so-called *sudoers*) and must specify what programs you can use in this mode. Luckily, on the typical Raspbian installation this has already been prepared, at least for the preconfigured `pi` user. On other Linux systems, such as the systems at work or school, you may not have the necessary privileges to use the `sudo` command.

Monitoring Hardware

Sometimes you need to check whether a peripheral device you have plugged in has been detected by Raspberry Pi. The easiest way to do that is to use the `lsusb` command (can you guess what the abbreviation stands for?), which shows all the detected USB peripherals. You can also check what the core of the system, called the *kernel*, has to say about the device. The kernel writes a kind of log, which you can read through the `dmesg` command—that is, *display messages*. Many messages will appear, which you can view in `less`:

```
pi@raspberrypi ~ $ dmesg | less
```

To find out what your CPU's capabilities are, use the command:

```
pi@raspberrypi ~ $ cat /proc/cpuinfo
```

You'll see a series of lines containing information about the processor(s). Look for the

Revision entry at the end and check its value: Raspberry Pi model As should display 0002 or 0003, while higher values (including characters, like 000f) identify B models and beyond.

The Graphical User Interface

Now that we have completed a quick overview of the use of the terminal, you can move on to the graphical environment. To start the graphical environment, type the `startx` command. The graphical environment, or *desktop environment*, is made out of two parts: a *graphics server* (the standard is called *X11*) and a *window manager*. To put things simply, the graphics server provides all the libraries and components to create the environment on which the window manager relies; the window manager creates a nice-looking and usable interface. One of the good things about Linux is that it can work on low-performing machines, since its window manager uses up few resources. The default window manager that Raspberry Pi uses is the *LXDE* desktop environment, shown in Figure 18-6.

Figure 18-6 *The LXDE desktop of Raspberry Pi*

The desktop should look familiar, at least in the basic sense that it looks a lot like other desktops you've seen on Windows, Mac, and Linux. At the top of the screen there's a menu with a number of icons, applets, and widgets. Starting from the left, you find the menu button: a small button with the Raspberry Pi logo and the word *Menu*. Then, you have shortcuts for a browser, file manager, Terminal (which lets you get to the shell even when you're in the graphical environment), Mathematica, and the Wolfram language environment. To the right is an applet that shows the current CPU usage, followed by the current time.

There is one icon on the desktop, the Trash.

Raspbian has some very useful applications already installed, such as an editor, utilities, programming tools, Internet programs, games, and system utilities. You can have a look at the installed programs by opening the menu: just click the bottom-left button. The Raspbian menu has a first level of main headings, under which you can see all the installed programs:

- The Programming menu contains the development environments for the Python language, Scratch, Wolfram, and others.
- The Internet menu contains the Pi Store (where you can find more software), a link to more Raspberry Pi resources, and the web browser.
- Under Games, you'll find Minecraft Pi and some games written in Python.
- Under Accessories there are many useful programs: a compressed file manager, the calculator, a file manager, an image and PDF viewer, your friend the terminal, a task manager to see what's running on your Pi, and a text editor.
- Preferences contains the programs that allow you to adjust the system settings. You can change elements of the graphic environment's appearance, and desktop, mouse, and keyboard behaviors. You can also configure your Wi-Fi connection here.
- The Run menu lets you type any Linux command and run it.

- To close the session or shut the system down, you would click Shutdown>...but don't do this yet!

Python

There's a snake hiding in Raspberry Pi: a python!

Well, not a real snake. In fact, the computer language Python is based on nothing more dangerous than the passion that its inventor, Guido van Rossum, has for Monty Python's Flying Circus. Van Rossum created this language to embrace simplicity and readability, without sacrificing the ability to create complex programs. With Python, writing programs for Raspberry Pi is fast and simple; moreover, it can be easily expanded with libraries, making it easy to access the GPIO pins and to control external hardware.

Teaching you everything about Python is not one of the goals of this book, but we are going to briefly show how to use it and what you can do with this powerful language. You'll see how different it is from other languages we analyze in this book, such as Processing or Arduino.

We have already said that the language you use with Arduino is a *compiled language*: you write the source code (or the sketch) in a more or less human-comprehensible language; after that, a program called a compiler turns it into *machine code*, a series of 0s and 1s that the computer processor can understand. Python, on the other hand, is an *interpreted language*; that is, a program called an *interpreter* reads each single instruction in Python and runs it straightaway, with no need to compile anything.

The Python interpreter is interactive: it behaves like a terminal where you can enter commands and get results right away. Launch the Python interpreter from the terminal, by typing `python` and pressing the Enter key.

You'll see some text lines appear, and a cursor waiting for your action. You can print "hello world" by typing:

```
print("hello world")
```

That wasn't hard, was it? The interpreter also works as a calculator, so if you write `2 + 3` and hit the Enter key, you get `5`.

To exit the interpreter, type `quit()` or press Control-D.

The instructions that you use interactively can be saved in a *script*—a text file that you'll get the interpreter to read (scripts are to Python as sketches are to Arduino). With the nano text editor, create a file named *helloworld.py* and write:

```
print("hello world!")
```

Save and close. Then, from the terminal, execute the script with:

```
python helloworld.py
```

There is your "hello world!" on screen.

In the graphical environment, you can choose Menu→Programming→Python 3. This is the Python development environment, and it's much more than a simple editor: you can use it to write programs, execute them, or *debug* them—that is, follow their execution step by step and "peek" into variables and code, in order to see what happens at every moment of the program so you can identify any problems.

As soon as you start it, the development environment displays the Python version number, and other information that you don't need right now. Open a new window with the New Window command, found under the File menu. Now you can type:

```
print("hello python!")
```

Next, press F5, and when prompted, save the code somewhere in a file called *hellopython.py*. Your code will be executed and the words "hello python!" will appear in the main window.

There are some differences between Python and the other languages in this book:

- In Python, there is no need to use a semicolon (;) after each statement.

- To define code blocks, you don't use curly braces ({}); in Python, you structure the code through formatting by using spaces and tabulating everything neatly.

You need to be very tidy and precise: each block must have the same indentation—that is, the same number of spaces—before each instruction. Here is some code with the wrong indentation:

```
num = 11
if (num > 10):
print (num)
     print ("is greater than 10"),
else:
     print (num),
     print ("is less than or equal to 10")
```

The `print` instruction in the second line is not aligned with the instructions in its block and Python will display an error. The correct format is the following:

```
num = 11
if (num > 10):
     print (num),
     print ("is greater than 10")
else:
     print (num),
     print ("is less than or equal to 10")
```

When you use the `print` statement without a comma at the end, it displays a line break after whatever it prints. If we didn't put a comma after `print (num)` the output of this script would be:

```
11 is greater than 10
```

You can create single-line comments by using the # character:

```
if (num > 10):
     #print the num value
     print (num),
```

Python has the same rules as other languages with regard to variable names: no exotic characters, spaces, or punctuation are allowed, and the name must not start with a number.

For further information about Python, there are many websites you can look up. The official documentation is at *https://www.python.org/doc/*.

GPIO

Depending on which model you have, a Raspberry Pi has between 17 (Models A and B) and 28 (A+, B+, and Raspberry Pi 2) general-purpose input/output hardware ports, in a connector strip of two rows of 13 or 20. In addition to the GPIO pins, which you can use as digital inputs or outputs, you'll also find some 3.3V and 5V power supplies and ground (GND) pins.

Raspberry Pi is way more delicate than Arduino, so you need to be very careful with the connections: its GPIO pins work only with 3.3V; if you tried to connect 5V, you could irreparably damage your Raspberry Pi.

Hello World

For this experiment, you are going to use an LED (choose one rated for 3.3 volts or more), a breadboard, and a few cables with male-female terminal connectors (Figure 18-7). Your goal is to try to switch the LED on and off by using some simple terminal commands.

Figure 18-7 *A male-female terminal connector to connect Raspberry Pi to components*

With your Raspberry Pi switched off, mount the LED on a breadboard and connect it, as shown in Figure 18-8, to the GPIO25 pin and to the ground pin (GND). The GPIO25 port puts out 3.3V, and your LED uses 3.3 volts, so you can get away without using a resistor for this quick test. For long-running projects, you're going to want a resistor between the power source and any LED you use.

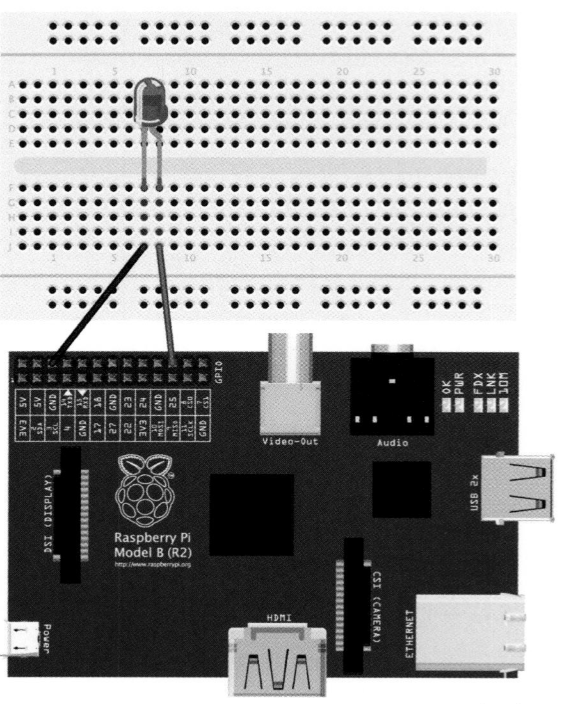

Figure 18-8 *Connect the LED to the GPIO25 pin of Raspberry Pi*

The Fritzing (http://fritzing.org) diagrams in this section show the Raspberry Pi A and B (26-pin) layout rather than the newer 40-pin layout. However, the first 26 pins of the 40-pin layout are identical to the original layout, so you can hook it up the same way on any model Pi.

Supply power to the Raspberry Pi and wait for it to boot up. Log in, but don't bother starting the graphical environment. If you did start it (with `startx` or if you've used `raspi-config` to configure it to automatically boot to the desktop), open a Terminal window.

When you change into a superuser shell with `sudo -s`*, your shell prompt will change appearance subtly. Instead of showing a $ at the end as it did in the preceding example, it will show a #, as in* `root@raspberrypi:/sys/class/gpio#`*. In the examples that follow, we'll abbreviate the shell prompt to either $ (for normal user) or # (for superuser).*

1. Change to a special directory created by the operating system to allow you to work on the GPIO pins directly:

   ```
   pi@raspberrypi ~ $
   cd /sys/class/gpio
   ```

2. As you saw earlier, you can inspect the directory contents with `ls`. Try that now.

3. In order to work with the GPIO pins, you need superuser privileges, so, before continuing, launch the `sudo -s` command:

   ```
   pi@raspberrypi /sys/class/gpio $
   sudo -s
   ```

   This lets you run all your commands as the superuser. Be sure to type `exit` when you're done with this example so you can exit superuser mode.

4. Now you can let your Raspberry Pi know that you need access to the GPIO25 pin by writing the pin number into the file named *export*. The `echo` command displays its arguments to standard output (your screen or the terminal window). You're using the redirection operator (>) to send that output into the file named *export*:

   ```
   # echo 25 > export
   ```

5. In response to this command, the system will generate a *gpio25* directory containing files that allow you to use the GPIO port. If needed, you can release the pin from use with this command (but don't type this now):

   ```
   # echo 25 > unexport
   ```

6. Keeping the pin exported for now, cd into the *gpio25* directory, then show its contents with `ls`:

   ```
   # cd gpio25
   # ls
   ```

   You'll find some special files within the directory: *active_low*, *direction*, *edge*, *power*, *subsystem*, *uevent*, and *value*.

7. You need to create a file named *direction* to set the pin as either an input or output, which is similar to Arduino's `pinMode`. To configure the pin as an output, write the word `out` to the file. To configure it as an input, you'd use `in`. For this, you'll use `echo` again:

   ```
   # echo out > direction
   ```

8. With Arduino, to switch an LED on and off you used the `digitalWrite` function with `HIGH` or `LOW` as a value; the equivalent on Raspberry Pi is writing the values 1 or 0 into a special file named *value*. So, to switch the LED on you'll write:

   ```
   # echo 1 > value
   ```

   whereas to switch it off:

   ```
   # echo 0 > value
   ```

When you're done switching the LED on and off, unexport it and exit out of superuser mode:

```
# cd /sys/class/gpio
# echo 25 > unexport
# exit
$ cd
$
```

A Flashing Python!

Luckily for you, there is a library in Python that can manage the GPIO pins, which makes everything very easy. This library is called RPi and should be already installed; if not, you can install it with:

```
$ sudo apt-get update
$ sudo apt-get install python-rpi.gpio
```

The library author has provided two ways of identifying Raspberry Pi pins: the first convention is *logical numbering* and can be selected in your Python code with:

```
GPIO.setmode(GPIO.BCM)
```

The second convention follows the *physical numbering* of the pins on the board and is activated with:

```
GPIO.setmode(GPIO.BOARD)
```

We'll use BCM numbering because it's what we were using in the preceding example when you blinked the pin, and it's commonly used in Raspberry Pi accessories and documentation (such as the diagrams in this chapter). If you counted the pin that you connected the LED to earlier, you may have wondered why you couldn't find a counting scheme that yielded the number 25. That's because *logical* pin 25 is connected to *physical* pin 22 on the header, as shown in Figure 18-9.

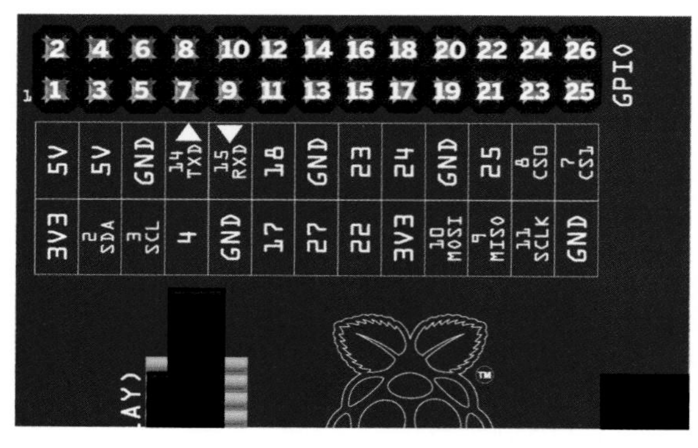

Figure 18-9 *Raspberry Pi pin numbers: physical (GPIO.BOARD) on top of the pins, logical (GPIO.BCM) on the labels below*

Also in Python, you need to configure the GPIO ports before using them. Use the `GPIO.setup` function followed by the pin number and the desired working mode (`OUT` for output and `IN` for input):

```
GPIO.setup(25, GPIO.OUT)
```

With pin 25 now set as an output, you are ready to switch the LED on or off with the `GPIO.output(port, value)` function. The GPIO library provides some constants for the port status: `GPIO.HIGH` and `GPIO.LOW`. So, you can switch the LED on with:

```
GPIO.output(25, GPIO.HIGH)
```

and switch it off with:

```
GPIO.output(25, GPIO.LOW)
```

Now you just need to create a loop that repeats these operations endlessly. Unlike Arduino, where you had the `loop` function, here you need to create the loop yourself. You'll use the `while` keyword, which executes instructions contained in the following code block until the condition you want to test no longer evaluates to true. If the condition simply consists of `True`, it will always be true and the cycle will never stop. Here is the start of your loop:

```
while True:
```

In Arduino you also had the `delay` function, which you could use to create pauses of milliseconds. In Python, you can use the `time` library and its related `sleep` function. While in Arduino delay times were measured in milliseconds, Python's `sleep` function uses seconds. To stop the script execution for two seconds, write:

```
time.sleep(2)
```

Remember that Raspberry Pi is not very precise in observing timing. This is because your program is being executed together with many others, so you have no guarantee that the operating system will observe the *exact* time you are setting.

Example 18-1 shows the complete listing of a blinking Hello World in Python.

Example 18-1 *"Blink" with Raspberry Pi*

```
# Hello World
import RPi.GPIO as GPIO
import time
pin = 25
# Use logical pin layout:
GPIO.setmode(GPIO.BCM)
GPIO.setup(pin,GPIO.OUT)
while True:
        GPIO.output(pin,GPIO.HIGH)
        time.sleep(1)
        GPIO.output(pin,GPIO.LOW)
        time.sleep(.5)
```

You can type this code into a file with the nano editor (`nano helloworldblink.py`), save it (Control-O, press Enter or Return, then quit with Control-X), and then execute it with superuser privileges (because you can't access the GPIO pins as a regular user):

Before you run the nano editor, make sure you're back in your home directory. You won't be able to save this file anywhere under /sys.

```
$ sudo python helloworldblink.py
```

Is the LED blinking?

Since the `while` loop doesn't really test for anything (`True` is always `True`, after all), this program will continue as long as the Raspberry Pi is powered. You have to forcefully stop it, by pressing the interrupt key combination, Control-C.

Button, Button

Try now to connect a button switch to the GPIO and read its status. First, switch the Raspberry Pi off by typing this command:

```
$ sudo halt
```

Once the power is disconnected, take a button and a 220 ohm resistor and connect everything as shown in Figure 18-10. Check all connections before switching the Raspberry Pi on.

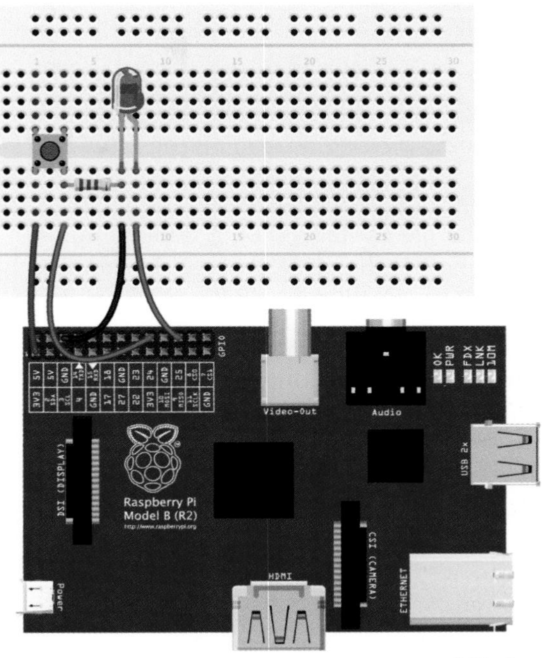

Figure 18-10 *Connecting a button to the GPIO24 pin and an LED to the GPIO25 pin*

1. As you did before with the output pin, try to read the status of a pin by using the terminal only. Move to the special directory */sys/class/gpio* and start a superuser shell:

   ```
   $ cd /sys/class/gpio
   $ sudo -s
   #
   ```

2. Let your Raspberry Pi know that you are going to work with the GPIO24 pin:

   ```
   # echo 24 > export
   ```

3. Next, change directory to the *gpio24* directory. If you run the `ls` command here, you'll now see the *gpio24* directory. Enter it and see what it contains:

   ```
   # cd gpio24
   # ls
   ```

4. You'll find the special files created by the Raspberry Pi. Next, create the *direction* file as shown to set the pin as an input:

```
# echo in > direction
```

5. You can detect whether the button is being pressed by reading the value file:

```
# cat value
```

   The file contains 0 or 1, depending on whether the pin is pressed. Try holding down the button while you type that command.

6. When you're done, unexport the pin, exit the root shell, and go back to your home directory:

```
# cd /sys/class/gpio
# echo 24 > unexport
# exit
$ cd
$
```

Let's do the same in Python. To use pin 24 as an input, the code is:

```
GPIO.setup(24, GPIO.IN)
```

You can observe the pin status with the `GPIO.input` function:

```
GPIO.input(24)
```

This will return `True` or `False`.

Example 18-2 shows the complete listing. Make sure you're back in your home directory, then use nano to save it in a file as you did with the previous example (for example, `nano button.py`); then, run it as you did before (for example, `sudo python button.py`):

Example 18-2 *Who pressed my button?*

```
import RPi.GPIO as GPIO
import time
# Use logical pin layout:
GPIO.setmode(GPIO.BCM)
button=24
led=25
GPIO.setup(led, GPIO.OUT)
GPIO.setup(button, GPIO.IN)
while True:
    if (GPIO.input(button) == True):
        GPIO.output(led, GPIO.HIGH)
        print("Ouch!")
        time.sleep(.1)
    else:
        GPIO.output(led, GPIO.LOW)
        time.sleep(.1)
```

After importing the GPIO module and defining the convention to number the pins, you declare pin 25 as the output and pin 24 as the input. With `while True`, you have created an never-ending cycle; within the cycle, you check whether anyone has pressed the button (`GPIO.input(24) == True`) and, if so, switch the LED on (`GPIO.output(25, GPIO.HIGH)`) and write "Ouch!" to the screen. After one-tenth of a second, the `while` loop resumes. If no one presses the button, the LED goes off.

If you look at the circuit, you will realize that the code is very similar to one of the sketches you wrote for the Arduino.

Arduino and the Raspberry Pi

Now that you know how both the Arduino and the Raspberry Pi work, you can use the two boards together to get the best out of each one. With a Raspberry Pi, it is very easy to use peripherals such as a webcam, which an Arduino would hardly be able to access, or to connect to and interact with the Internet. On the other hand, you could use Arduino to acquire analog data, which you can't do easily with the Raspberry Pi.

Even though you can install the development environment of Arduino on Raspberry Pi, you probably shouldn't, because the Arduino editor and compiler require a lot of computing power. The best approach would be to program Arduino with your computer and then connect it to the Raspberry Pi with the sketch already loaded.

Let's use an easy example of how to make the two boards communicate with each other: you'll switch on an LED connected to the Arduino by pressing a button on the Raspberry Pi keyboard. Not exactly rocket science, but it is a starting point.

First, on the breadboard connect an LED to a 220 ohm resistor, then connect that to Arduino pin 13 and Arduino GND.

Connect the Arduino to a computer and load the sketch shown in Example 18-3.

Example 18-3 *Switching the light on and off from the serial port (Arduino)*

```
void setup()
{
    Serial.begin(9600);
    pinMode(13, OUTPUT);
}
void loop()
{
    int n = Serial.read();
    if ((char)n == 'a')
    {
        digitalWrite(13, HIGH);
    }
    if ((char)n == 's')
    {
        digitalWrite(13, LOW);
    }
}
```

It's rather simple: in the `setup` section, you configure Arduino pin 13 as an output, and prepare the serial port. The loop section reads what comes from the serial port, and tries to understand what it has received by checking the characters against two expected values. If it has received an `'a'`, it switches the LED on pin 13 on; if it gets an `'s'`, it switches the LED off.

Now, try to find out which serial port you need to use. Connect the Arduino to the Raspberry Pi with a USB cable just as you would connect the Arduino to a computer.

Right after you plug it in, type the command `dmesg` from the terminal; in the last lines of the output, you can see that your Arduino has been detected, and that its serial port is ttyACM0 (in which case, the filename will be */dev/ttyACM0*):

```
[ 5307.783795] usb 1-1.4: New USB device
strings: Mfr=1, Product=2, SerialNumber=220

[ 5307.783813] usb 1-1.4: Manufacturer:
Arduino (www.arduino.cc)

[ 5307.783830] usb 1-1.4: SerialNumber:
74937303936351014261

[ 5307.826464] cdc_acm 1-1.4:1.0: ttyACM0:
USB ACM device
```

If the port is anything other than `/dev/ttyACM0`, replace the port filename in line 2 of Example 18-4 with the correct port filename.

On the Raspberry Pi, you'll write a program in Python to read what you type on the keyboard and send it over to Arduino, which will decide what to do with the characters it receives. In order to use the serial ports, you import the serial module, similar to Arduino's Serial library. Use `Arduino = serial.Serial(port, 9600)` to create an Arduino variable. Before working with the serial port, clean it with `flushInput`, in case there is some undesired character still in the queue. The `while` cycle keeps going until you press Q. In order to read from the keyboard, you use the `raw_input` function; each string is then printed onto the screen and sent to the Arduino with `Arduino.write(str)`.

Example 18-4 shows the complete listing in Python. Save it in a file as you did for the other Python examples, and run it with `python` *file name*. Because you're not accessing the GPIO pins, you don't need to use `sudo` to run this program.

Example 18-4 *Switching the light on and off from the serial port (Raspberry Pi)*

```
import serial
port = "/dev/ttyACM0"
Arduino = serial.Serial(port, 9600)
Arduino.flushInput()

while True:
   str = raw_input(">:")
   if (str == "q"):
      break
   print str
   Arduino.write(str)
```

If there's anything that doesn't work, it might be because the serial module has not been installed. To check that, open a terminal and, in the Python interpreter, write:

```
import serial
```

If any error comes up, it is possible that the module hasn't been installed. To install it, go to the Raspberry Pi terminal and type:

```
$ sudo apt-get install python-serial
python3-serial
```

Let's close this chapter with an exercise:

- Modify the preceding Raspberry Pi and Arduino code to have the LED on Arduino flashing at increasing or decreasing speed when you press the < (less than) and > (greater than) keys.

Processing

19

After many chapters on hardware, let's go back to software for a moment. Later in this chapter you'll see how to combine hardware and software again.

Processing is an open source programming environment that lets you easily create images, animations, and interactions. Created as a teaching tool, it is now used by thousands of students, researchers, artists, hobbyists, designers, professionals, and people in many other categories.

This development environment is very similar to the Arduino's, since Arduino's IDE was derived from Processing, so we're not going to recap the meaning of all the buttons and menu entries. However, the Processing language itself is slightly different, and you are going to see how it can get you started with *object-oriented programming* (OOP). Let's start with the basics.

Your First Sketch

Like Arduino, Processing's programs are called *sketches*, emphasizing their ephemeral nature. The simplest sketch is empty and barely does anything: it only opens a new window called a *display*. Download Processing from *https://processing.org/download/* and install as directed on the website. Run Processing, and an empty sketch will appear. Next, click the Run icon (the leftmost icon in the main window), as shown in Figure 19-1.

Figure 19-1 *An empty sketch opens the display window*

Let's get it to write the words "Hello World!" on the console (the lower part of the Processing IDE, as shown in Figure 19-2):

```
print("Hello World!");
```

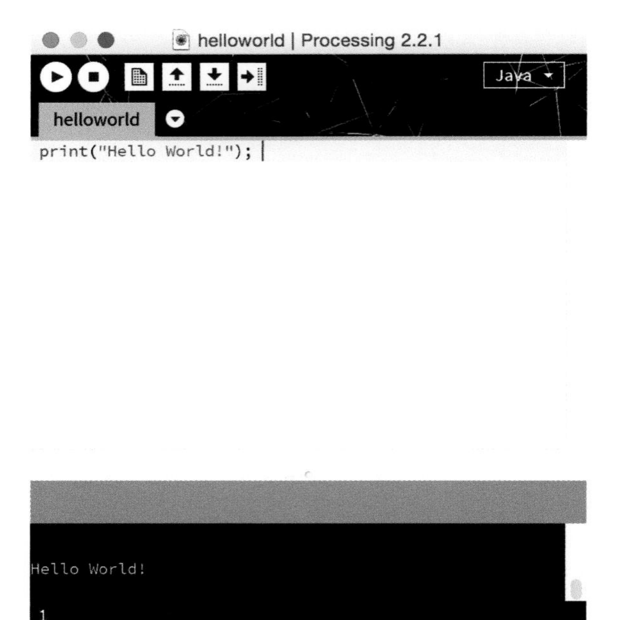

Figure 19-2 *A greeting to all from the Processing console*

To make your Hello World more appealing, you can do more and try to write on the display. The instructions are slightly more complex, yet it's not too bad:

```
fill(0);
text("Hello World", 20, 20);
```

The `fill` function defines what color you are going to use for the writing—in this case zero, or black. The `text` function writes the message you put in quotes as the first parameter, while the second and third parameters specify the position where the writing is going to start from. The position of any point on the display is defined by the pair of *x* and *y* coordinates, similar to the Cartesian plane you learned about in school, shown in Figure 19-3.

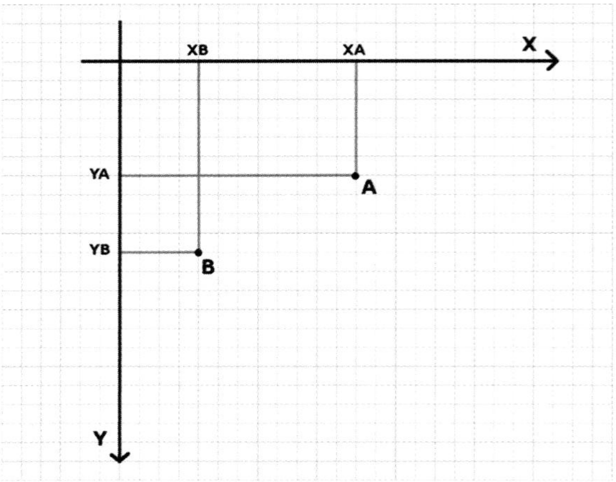

Figure 19-3 *The coordinates system used in Processing*

As you can see, the y-axis points down, unlike standard mathematical convention. It might be hard to remember the difference, but just like any other convention, you get used to it. You could add one further coordinate, the position on the z-axis to be used in 3D representations.

Let's try to execute your sketch (Figure 19-4).

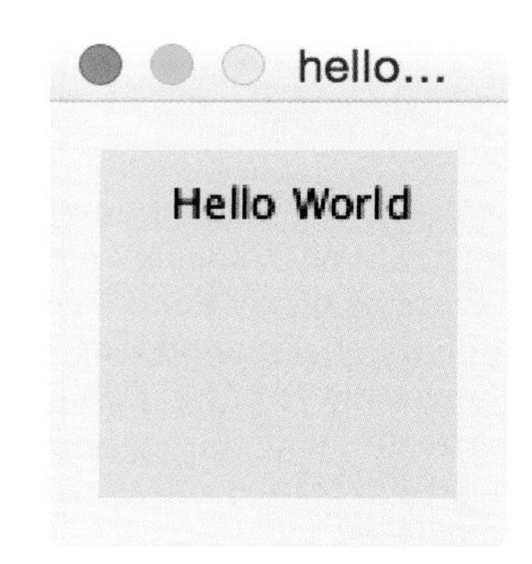

Figure 19-4 *A "Hello World!" on your display*

Since Processing is widely used for visual representations, you should now try to draw a line. It's a very simple instruction:

```
line(0, 0, 50, 50);
```

The first two parameters are the coordinates of the first point, the second two of the second. When you run it, it should look like Figure 19-5.

Figure 19-5 *Drawing a line*

If you want a comprehensive manual on all the functions, you can click Help on the menu, then Reference: a browser window will open, where you will find all the information you need. With this information and what you have learned in the previous chapters, you shouldn't have any problem.

Let's now try to make things more complicated by introducing a loop. You already know that if you have to repeat a series of commands a number of times, rather than duplicating the instructions you can use the `for` statement. Let's try to reproduce in Processing a pattern you might have drawn a thousand times in a technical drawing or drafting class, shown in Figure 19-6.

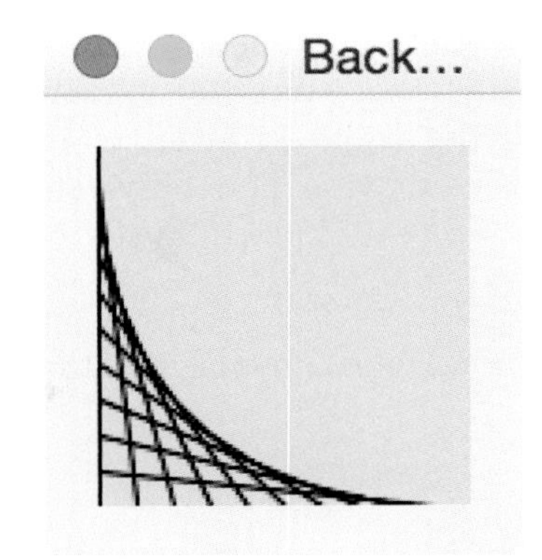

Figure 19-6 *Back to school*

This looks complicated, right? It is actually rather simple. Start by planning the algorithm, which is the series of instructions you are going to use. You want to connect two points: the first one starts at the (0, 0) coordinates and moves along the y-axis; the second one starts at (0, `height`) and moves parallel to the x-axis. The term *height* refers to the display height; in the same way, you talk about its *width*.

You are using four variables to define the coordinates of the two points. The first part of this loop is:

```
int x1 = 0, y1 = 0, x2 = 0, y2 = height;
```

Notice that you have set variable `y2` not to a number, but to the word `height`. Processing has a number of parameters called *system variables*. These are variables that Processing uses to tell you about its internal state. In this case, `height` contains the height in pixels of the default window.

Next, move the first point down and the second one to the right:

```
x2 = x2 + 10, y1 = y1 + 10;
```

Repeat the cycle until the second point reaches the extreme right of the display (or until the first point reaches the bottom). So, the condition is:

```
x2 < width;
```

Notice that you are comparing x2 to something called width? That's another system variable, containing the width in pixels of the default display. Now draw a line between the two points:

```
line(x1, y1, x2, y2);
```

Since x1 and y2 won't change, you can write:

```
int x1 = 0;
int y2 = height;
for (int y1 = 0, x2 = 0;
     x2 < width;
     x2 = x2 + 10, y1 = y1 + 10)
{
  line(x1, y1, x2, y2);
}
```

Type this sketch into Processing and see what you get. Notice how complex the for loop is. As you see, it's possible to use a for loop to change two variables simultaneously, both x and y.

Let's Get a Move On!

With a few code lines you have created a rather complex, yet static, drawing. Let's move on to something animated: you are going to draw a circle and move it back and forth on the screen. You need to break up the problem into discrete steps. First, you are going to draw a circle, then you'll see how to move it to the right, and, finally, you'll learn how to make it go back to the left. But first, get comfortable and increase the size of the display with the following instruction:

```
size(500, 200);
```

Next, draw a circle, centered vertically on the display: use the ellipse command, which accepts the center coordinates and the dimensions of the two axes as parameters:

```
ellipse(0, height/2, 50, 50);
```

Now you can move it right until the center reaches the edge. You can use a simple loop within the draw block. Enter the following into a new Processing sketch:

```
size(500, 200);
for (int x = 0; x < width; x++)
{
   ellipse(x, height/2, 50, 50);
}
```

Now, launch the sketch. The outcome is not exactly what you expected. You see a bunch of curved lines spread across the screen from left to right, ending with a half-circle of white hugging the right edge of the display (Figure 19-7).

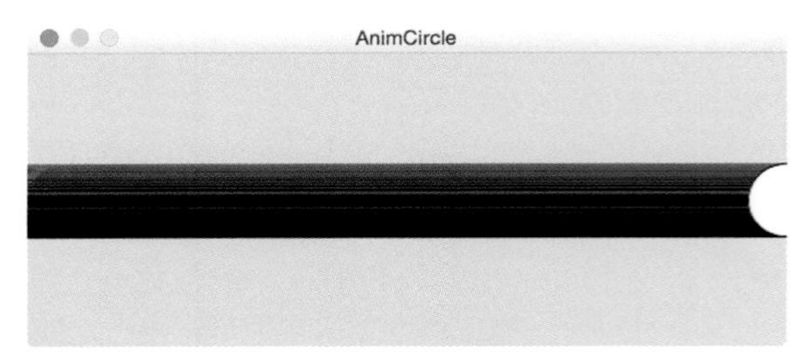

Figure 19-7 *Where is your animation?*

What if you tried to clear the screen before drawing each circle?

```
size(500, 200);
for (int x = 0; x < width; x++)
{
   background(220);
   ellipse(x, height/2, 50, 50);
}
```

The background instruction fills the whole screen with a specific color, in this case pale grey (see Figure 19-8). (You'll get a more in-depth look at colors later in this chapter.) Is it better? A little, because you don't see all those curved lines anymore. (As you might have already figured out, those curved lines were the previous circles that had been written on the display.) But…what about the animation? Why can't you see the circle moving?

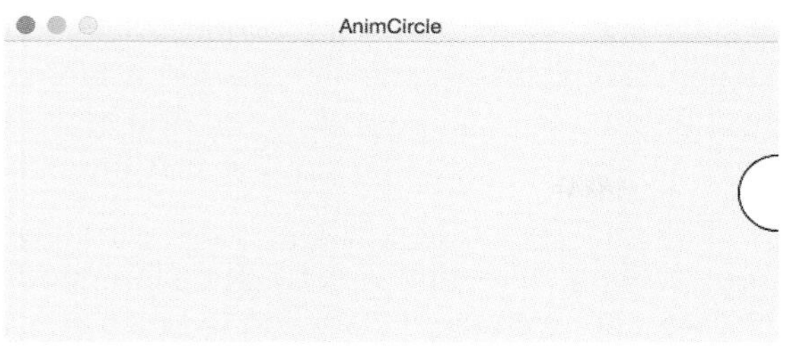

Figure 19-8 *A bit better, yet still not there*

The answer is simple: the loop operates too fast for you to see. As in Arduino, the canonical form of the Processing sketch structure is made of two blocks: the first one is called *setup*, while the repeated block is called *draw*, instead of *loop* as in Arduino. Why *draw*? Not just to be different: the premise behind Processing is that everything you create in this visual language will be displayed on some kind of device. Unlike Arduino, however, the setup and loop/draw block is not absolutely mandatory in Processing.

You can decide what size you want your display to be, the color of its background, and at what speed it is going to perform your instructions; all this information falls into the setup block.

In order to see the animation, you should draw the circle just once for each execution of the `draw` method (another name for a function). Start a new sketch. First, you need to declare the `x` variable, which must be outside of the draw block so that it will not be reset each time. Then, you declare the size of the display in the `setup` method. Then, instead of a `for` loop, you have a `draw` method, which includes the necessary `x` increment:

```
int x = 0;
void setup()
{
   size(500,200);
}
void draw()
{
   background(220);
   ellipse(x, height/2, 50, 50);
   x++;
}
```

Finally, you can see the animation! (But you'll have to take our word for it in Figure 19-9.)

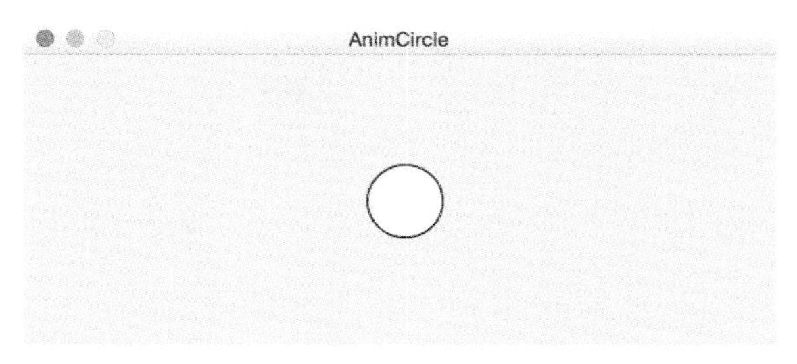

Figure 19-9 *And yet it moves!*

How can you make the circle go back now? You made the circle move to the right by increasing `x` by a certain amount; to move the circle to the left, you should just decrease `x` by the same amount. But how do you know which direction you're going? And how do you know when it is time to switch? And how can you change the value of `x`?

First, you set a variable to tell the direction. You call this variable `pitch` and write its declaration outside of `draw` and `setup`, as you have done with `x`:

```
int pitch = 1;
```

The increment instruction is:

```
x = x + pitch;
```

The points where you change direction (i.e., where you change sign before the `pitch` variable) are the windows' edges:

```
if (x == 0 || x == width)
{
  // change direction
}
```

The two || (double *pipe*) characters are a *Boolean operator* called *or*: it returns a true value if either the first condition (`x == 0`) OR the second condition (`x == width`) is true. If either condition is true, processing assesses the whole expression in brackets as true. You'll tell it to then change the sign of the `pitch` variable, causing it and ultimately the circle to change direction:

```
if (x == 0 || x == width)
{
   pitch = -pitch;
}
```

The complete sketch code is shown in Example 19-1.

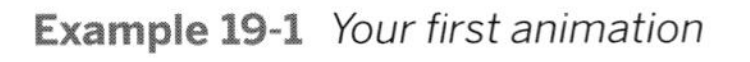

Example 19-1 *Your first animation*

```
int x = 0;
int pitch = 1;
void setup()
{
   size(500, 200);
}
void draw()
{
   background(220);
   ellipse(x, height/2, 50, 50);
   x = x + pitch;
   if (x == 0 || x == width)
   {
      pitch = - pitch;
   }
}
```

To practice more, try to modify the code to change the speed of the circle. It's not hard!

How Many Circles?

What if you wanted to draw and move many circles, defining each circle's speed and color? This is where things get a little complicated. For a single circle, you need two coordinates for its center, two values for its height and width (you actually draw ellipses, so you need two parameters even though they are the same), and a variable for the speed and one for the color; so, you are talking 6 variables for each circle. If you wanted 10 circles, you would have to manage 60 different variables. Ugh!

In this context, *arrays* come to the rescue: these are data structures that can contain many items of the same kind. It's like having a chest of drawers: there's something in the first drawer, something else in the second, another something else in the third, and so on for each drawer, but all of them are contained in the same chest. Each place (each drawer) is identified with a number, called an *index*. And as with most things computer-based, the first index is not 1, but 0.

You can only have one variable for each "kind" of data, and this variable is going to be an array. You will have an array for the *x* coordinate of the circle centers, one for the *y* coordinate, one for the speed, and so on. As with a chest of drawers, you have to decide beforehand how many indices an array is going to have. So if you want 10 circles, you'll write:

```
int numOfCircles = 10;
int[] x = new int[numOfCircles];
int[] y = new int[numOfCircles];
int[] dim = new int[numOfCircles];
int[] colour = new int[numOfCircles];
int[] pitch = new int[numOfCircles];
```

Colour Versus Color

You've probably already noticed that the color array is spelled *colour*, as if you were members of the British Commonwealth. Did you steal this Processing sketch from the Queen of England?

Of course not. The problem is that `color` is a protected Processing system variable, which you're not allowed to use as your own variable. You can tell because in the Processing IDE, the word `color` is literally a different color than the other words; that indicates a restricted word. By calling the array *colour* you avoid the restriction problem, while also making it clear what the array controls.

To create an array, you need to specify the kind of data that the drawers are going to contain, followed by square brackets before the name, and its length in the second part of the instruction. You also need to use the keyword `new` because you are creating a peculiar variable: an *object*.

The centers of the circles will start at 0 on the x-axis, and will be spaced equally throughout the y-axis:

```
int distance = height / numOfCircles;
int temp = distance / 2;
for (int i = 0; i < numOfCircles; i++)
{
   y[i] = temp;
   temp = temp + distance;
}
```

To decide the size and color of every circle, you'll use the `random` function, which gives you random numbers (*pseudorandom* actually, because even though they seem to be random, they are generated by an algorithm) between the two numbers given as parameters. The random function renders *float* values (i.e., decimal numbers), but you need whole values; to convert them, use the `int` keyword:

```
dim[i] = int(random(5, 2.5 * distance));
```

Do the same with colors. For now, you can stick to greyscale:

```
colour[i] = int(random(255));
```

Finally, the pitch:

```
pitch[i] = int(random(1, 15));
```

Because the pitch is no longer constantly 1 or −1, certain circles might not necessarily reach the exact edge of the display before returning. A small circle with a large pitch could disappear to the left or the right before it reverts:

```
if (x[i] > width || x[i] < 0)
{
   pitch[i] = - pitch[i];
}
```

Example 19-2 shows the complete listing, and you can see the sketch in action in Figure 19-10.

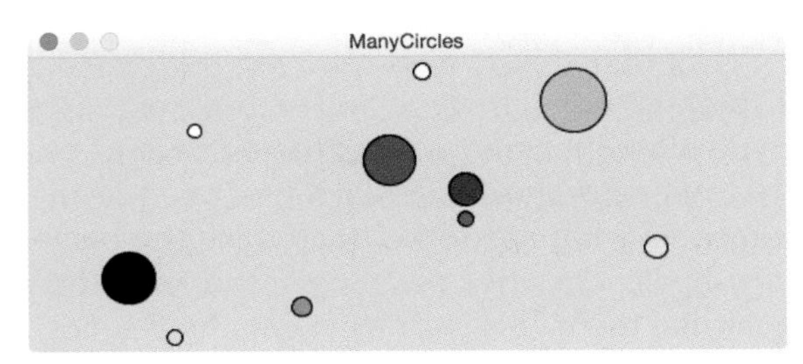

Figure 19-10 *Your circles in motion*

Example 19-2 *As many circles as you can*

```
int numberOfCircles = 10;
int[] x = new int[numberOfCircles];
int[] y = new int[numberOfCircles];
int[] dim = new int[numberOfCircles];
int[] colour = new int[numberOfCircles];
int[] pitch = new int[numberOfCircles];

void setup()
{
   size(500, 200);
   int distance = height / numberOfCircles;
   int temp = distance / 2;
   for (int i = 0; i < numberOfCircles;
      i++)
   {
      y[i] = temp;
      temp = temp + distance;
      dim[i] = int(random(5, 2.5 *
            distance));
      colour[i] = int(random(255));
      pitch[i] = int(random(1, 15));
   }
}

void draw()
{
   background(200);
   for (int i = 0; i < numberOfCircles; i+
+)
      {
      x[i] = x[i] + pitch[i];
      fill(colour[i]);
      ellipse(x[i], y[i], dim[i], dim[i]);
      if (x[i] > width || x[i] < 0) {
      pitch[i] = - pitch[i];
      }
   }
}
```

What if you wanted colored circles (Figure 19-11)? As we mentioned, Processing puts at your disposal the `color` variable, for which you can use the three components *RGB* (red, green, and blue). Change the sketch by modifying the color array:

```
color[] colour = new color[numberOfCircles];
```

Next, initialize it with random values. Change the definition of `color[i]` in `setup` to the following to get a result similar to Figure 19-11:

```
colour[i] = color(int(random(255)),
int(random(255)), int(random(255)));
```

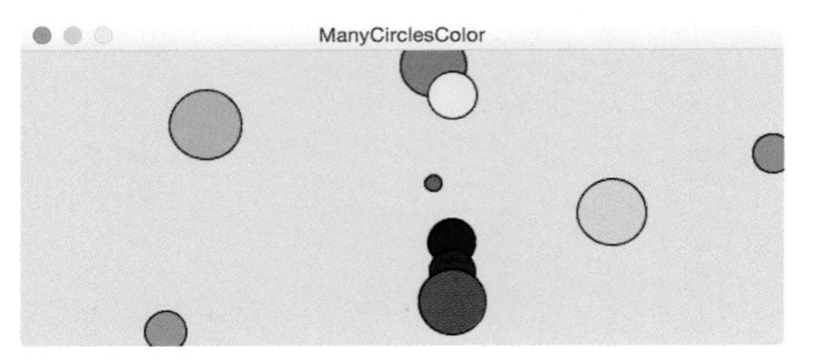

Figure 19-11 *You can color all circles at once*

See how easy that was? Thanks to the power of arrays, you can modify the color of 10 circles at once. If you wanted to, you could increase the number of circles to nearly any number you wanted, just by modifying a single line in your sketch!

I've Got the Power!

Now that you understand the basics, you can try to interact with the computer. Processing is capable of reacting to the mouse position and mouse clicks, to keypresses, and to other events. For instance, let's try to draw a dot wherever you click the mouse. The position of the mouse pointer is identified with the `mouseX` and `mouseY` system variables. To indicate when you are pressing the mouse button, use the `mousePressed` system variable, which can take the `true` or `false` values. The sketch is very simple; you just need to put it into a draw block:

```
void draw()
{
   if (mousePressed)
   {
      point(mouseX, mouseY);
   }
}
```

There isn't much to see, right? Increase the dot size by putting this at the top of the draw block:

```
strokeWeight(10);
```

As you can guess, by increasing the parameter, you increase the size of the dot. Alternately, you could put the `strokeWeight` instruction in `setup`.

You can also use the keyboard to give Processing some information. Each time a key is pressed on the keyboard, that generates an *event*, which Processing knows about. You will use the `keyPressed` method to see if a keypress event has occurred.

Once you are within the `keyPressed` method, you can use the `key` system variable to find out *which* key has been pressed. If, for instance, you wanted to print to the console everything you type, you could use this simple sketch (you must have a `draw` function, even an empty one, in order for `keyPressed` to work):

```
void keyPressed()
{
   print(key);
}

void draw() {
}
```

Programming with Cartoons

In cartoons it is very common to find worlds in which objects speak and act human. In such fantastic worlds you might hear dialogues such as the following:

Dog: "Home, why are you sad?"

Home: "Because today is such a beautiful day and I am all dark...so I am too hot!"

Dog: "No problem: paint yourself white!"

Home: "Great idea!"

What's all this got to do with Processing? This type of world demonstrates the essence of object-oriented programming: it is not you saying precisely what has to be done (take the paintbrush, dip it in white paint, paint the walls, then wait for them to dry and repeat three times), but rather you telling the objects to do something, and giving them the responsibility of managing the process themselves: "Home, paint yourself white!"

Let's try to understand the difference. You want to simulate a surface (like a body of water) onto which water drops fall, creating rings that become larger little by little and eventually disappear. Start with a single drop, as shown in Example 19-3.

Example 19-3 *Water drop simulation*

```
int bg_color = 200;
int x;
int y;
int rayMinimum = 10;
int rayMaximum = 100;
int ray = 10;
int initialColor = 100;
int colour = initialColor;

void setup() {
  size(200, 200);
  x = width/2;
  y = height/2;
  fill(bg_color);
}

void draw() {
  background(bg_color);
  stroke(colour);
  ellipse(x, y, ray, ray);
  ray = ray + 1;
  colour = colour + 1;
  if (ray > rayMaximum) {
    ray = rayMinimum;
    colour = initialColor;
  }
}
```

So far there is nothing new here; we're using the same techniques you saw earlier. What you would like, now, is to create a drop that you can instruct like some anthropomorphic cartoon character:

```
void draw() {
  background(bg_color);
  drop.update();
  drop.draw();
}
```

As you can see, the code is much cleaner and more comprehensible: given a drop, you just need to ask it to update and draw itself.

Classes and Objects

Of course, first you have to create this drop and then insert into it all instructions you had in the droplet sketch.

The drop is an *object*, which means it is characterized by a series of operations (in your case, `update` and `draw`) and by some properties (e.g., positions, speeds, dimensions, and colors), which keep track of the effects of such operations. It is correct to assume that all drops behave in the same way (at least in object-oriented programming it works this way). So you can say that all drops are defined by a sort of blueprint that says how they are made and how they behave. Such a blueprint is called a *class*. In particular, your class will be called `Drop`, and the drops are called *instances* of the `Drop` class.

Before you do anything, save your sketch as `DropTest`. Next, create a new class by clicking on the arrow on the right side of your programming environment, then click on the New Tab menu (Figure 19-12). Processing will ask you to name your file and you will type Drop, as shown in Figure 19-13. Click OK, and your job is done!

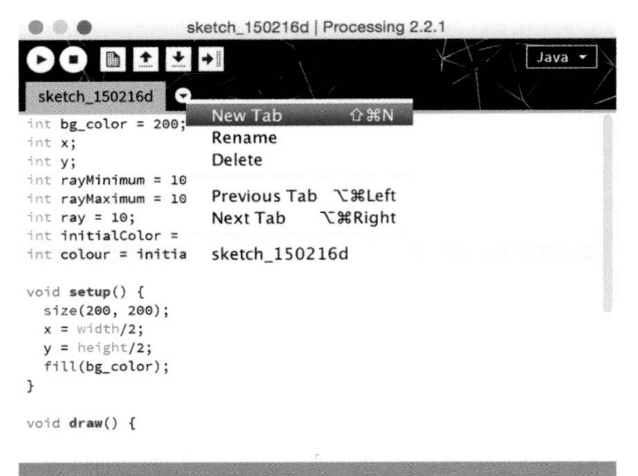

Figure 19-12 *Adding a new file to your project*

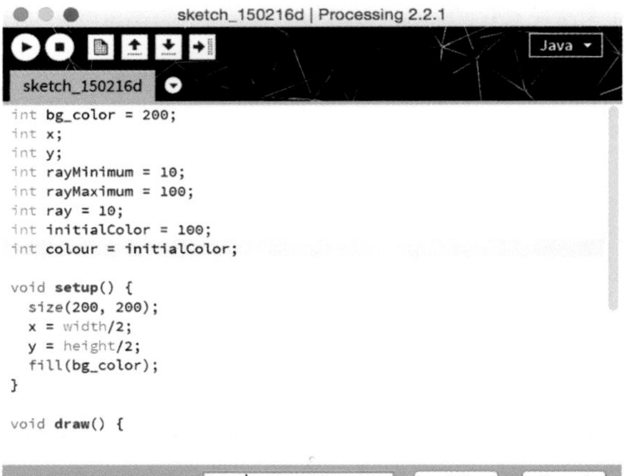

Figure 19-13 *Choosing a name for the new file*

Now that the file is created, define the class with the following. Everything else you type as part of the class will go inside Drop's curly braces, { and }:

```
class Drop {
}
```

In the Drop class, include all information you need (Figure 19-14):

```
int x;
int y;
int rayMinimum = 10;
int rayMaximum = 100;
int ray = 10;
int initialColor = 100;
int colour = initialColor;
```

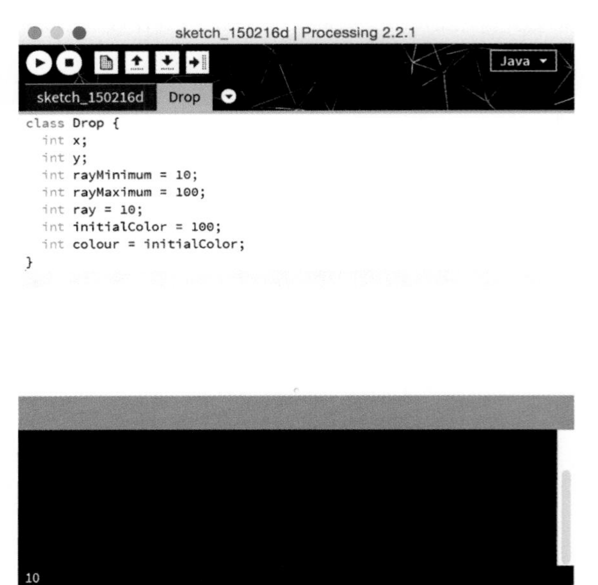

Figure 19-14 *A new home for your Drop*

I Want...a New One!

When you create a new drop you have to decide where its center is, what its minimum radius (the variable ray) and maximum radius are, and what its initial color is. To create an object, you use a special method called a *constructor*, whose name (in Processing, and some other languages like C++) is always the same as the class name. In the constructor the parameters of the object are usually initialized:

```
Drop(int x, int y, int rMin, int rMax, int
col){
  this.x = x;
```

```
    this.y = y;
    this.rayMinimum = rMin;
    this.rayMaximum = rMax;
    this.ray = rMin;
    this.initialColor = col;
    this.colour = col;
  }
```

The word `this` refers to the object itself, so the instruction `this.rayMinimum = rMin;` means "take the `rayMinimum` variable of the drop object you are working on and set it to the value of the `rMin` parameter."

OK, but What Should I Do with It?

In your main sketch you will ask the drop to update and draw itself. Write the two methods, using the code you have removed from the other part:

```
  void update(){
    ray = ray + 1;
    colour = colour + 1;
    if (ray > rayMaximum) {
      ray = rayMinimum;
      colour = initialColor;
    }
  }

  void draw(){
    stroke(colour);
    ellipse(x, y, ray, ray);
  }
```

If you read that carefully, you'll notice that you didn't use the `this` keyword. This is because it is optional most of the time; in the constructor you used it because two of the arguments that it got passed (x and y) had the same names as two properties of your `Drop` class, so you needed to be explicit for these assignments; the others were simply kept consistent.

The `update` and `draw` methods do not give you anything back as return values when you call them, so before them you write `void`. Done! Your `Drop` class is completed. Now you have to use it.

Using a Drop

Let's go back to your main sketch. To create a new drop, and in general a new object, you have to use the `new` command followed by the object's name, with the necessary parameters, as you have already seen in the examples with arrays. Before the sketch, you have to declare that you want to use an object you will call `drop` of the `Drop` type:

```
Drop drop;
```

In the setup, you just need to add the instruction:

```
drop = new Drop(width/2, height/2,
                     rayMinimum,
rayMaximum,
                     initialColor);
```

and...there you go! Let's try to execute the sketch; you will get the same result, but with cleaner and more manageable code, as shown in Example 19-4.

Example 19-4 *The sketch (DropTest) needed to use the Drop class*

```
int  bg_color = 200;
int rayMinimum = 10;
int rayMaximum = 100;
int ray = 10;
int initialColor = 100;
Drop drop;

void setup() {
  size(200, 200);
  fill(bg_color);
  drop = new Drop(width/2, height/2,
                  rayMinimum, rayMaximum,
                  initialColor);
}

void draw() {
  background(bg_color);
  drop.update();
  drop.draw();
}
```

Example 19-5 *The Drop class*

```
class Drop {
  int x;
  int y;
  int rayMinimum = 10;
  int rayMaximum = 100;
  int ray = 10;
  int initialColor = 100;
  int colour = initialColor;
  Drop(int x, int y, int rMin, int rMax, int col) {
    this.x = x;
    this.y = y;
    this.rayMinimum = rMin;
    this.rayMaximum = rMax;
    this.ray = rMin;
    this.initialColor = col;
    this.colour = col;
  }
  void update() {
    ray = ray + 1;
    colour = colour + 1;
    if (ray > rayMaximum) {
      ray= rayMinimum;
      colour = initialColor;
    }
  }

  void draw() {
    stroke(colour);
    ellipse(x, y, ray, ray);
  }
}
```

Raindrops Keep Fallin' on My Head...

What if you wanted many drops? For example, you want a new drop to fall any time you click on the mouse. In the main sketch you will need to keep track of all drops, but you don't know how many drops you are going to get, so the array won't work. A structure called `ArrayList` is needed: it is similar to the array, but it also allows you to have a variable number of elements. To create a list of drops, replace the line `Drop drop;` at the top of the `DropTest` sketch with the following:

```
ArrayList<Drop> drops = new
ArrayList<Drop>();
```

So, when you define the `drops` variable, you are not only saying that it is of an `ArrayList` type, but you are also indicating that the list will contain objects of the `Drop` type. The `ArrayList` is a class, so it has methods you can call.

The first method that interests you is the `add` method, which allows you to add a drop to the list: `drops.add(drop)`.

With the `size` method, you can specify how many drops you want and use a traditional `for` loop to draw all the drops. Change `DropTest`'s `draw` method to the following:

```
void draw() {
  background(bg_color);
  for (int i = 0; i < drops.size(); i++){
```

```
        Drop drop = drops.get(i);
        drop.update();
        drop.draw();
      }
    }
```

To get the object in the *n*th position you use the `get` method, using its position in the list as a parameter; just like with arrays, the first element of the list has an index of zero.

To add a circle to each click of the mouse you can use the `mousePressed` method, which reacts to the "mouse button has been clicked" event:

```
void mousePressed(){
 Drop drop = new Drop(mouseX, mouseY,
     rayMinimum, rayMaximum,
     initialColor);
 drops.add(drop);
}
```

Modify the `setup` method to appear as follows (you no longer need to define a `Drop` object there):

```
void setup() {
  size(200, 200);
  fill(bg_color);
}
```

Launch the sketch and try to click a couple of times: can you see the power you have at your disposal?

If someone creates a class and makes it available, it is very easy for you to write a sketch that uses it without your knowing how it is made. The only thing you need to know is how to use it. It's a bit like going to a restaurant and ordering a meal: you don't need to know how it is cooked, you just need to ask for it from the person taking your order (an instance of `Waiter`) and then you can simply devour it. In turn, the `Waiter` will not cook it directly, but rather he will ask someone else (a `cook` object of the `Chef` type) to prepare it. So, once you have ordered it from the `Waiter`, the meal will arrive at your table ready to be eaten, and this is all you need to know. In the same way you can build extremely complex systems by making many small, simple-to-manage objects interact with one another: instead of writing a very complicated sketch you will benefit from collaboration among objects, each one with its specific responsibilities whose complexities are hidden to you. Isn't it great?

Let's now turn our attention to the `Drop` class, and make some changes. To improve the effect you get when the circles overlap, you're better off leaving them empty:

```
void draw(){
  noFill();
  stroke(colour);
  ellipse(x, y, ray, ray);
}
```

After you click many times, the sketch gets very cluttered. So why don't you modify the `Drop` class so that, once the `rayMaximum` is reached, the drop disappears? This is the way rain works in the real world...so how can you do it? As always, there are different ways to reach your goal. How do you know if the drop has reached its largest dimension? Let's ask the drop: who would know better than it? However, you don't know the internal functioning of the drop, so you can't simply ask it, "Have you reached your largest dimension?" because you don't know if it is able to tell you; you can imagine this, but you don't know for sure. This principle is called *information hiding*, and following it helps you write better sketches. Now, let's have the drop use a Boolean method—that is, a method that returns one of the two possible values, true or false—called `disappear`:

```
boolean disappear() {
 return ray > rayMaximum;
}
```

We're not interested in increasing the drop's dimensions beyond its maximum value, so we can trim down `Drop`'s `update` method to:

```
void update(){
  ray = ray + 1;
  colour = colour + 1;
}
```

If the drop is no longer visible (because it has reached its largest dimension or for some other reason you are not aware of), it doesn't interest you anymore. Let's go back to the sketch that uses the `Drop` catch and modify its `draw` method to remove the drop from the `drops` list. Change the `for` loop as shown here (the result is shown in Figure 19-15):

```
for (int i = 0; i < drops.size(); i++){
  Drop drop = drops.get(i);
  if (drop.disappear()){
    drops.remove(drop);
  } else {
    drop.update();
    drop.draw();
  }
}
```

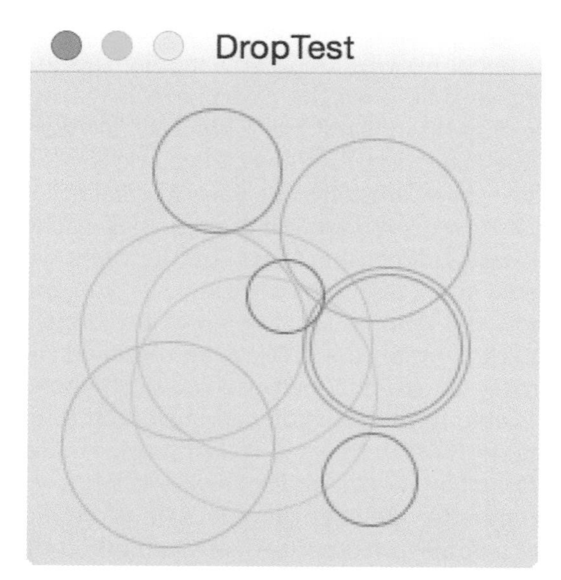

Figure 19-15 *Your rain animation is completed*

Processing, Meet Arduino!

Many artists use Processing for interactive installations, gathering information from the environment. Why can't you do it too? Let's write a sketch that interacts with Arduino. The communication channel we'll use is the serial port, so suppose you write a very simple Arduino sketch that slowly counts from 1 to 255, and then down again, all the while sending that number to the serial port. What you have to do in Processing is to read this data and use them for your purposes—for example, by drawing a circle with its radius proportional to the value sent by the Arduino.

Libraries

As with Python, Processing is not able to read the data from the serial port and needs to use an external library. To use a library, you first have to *import* it (i.e., tell Processing you want to use it):

```
import processing.serial.*;
```

The processing.serial library contains a class called `Serial`, which represents a serial port. To use it, you first have to declare it:

```
Serial SerialPort;
```

To build a new object of the `Serial` type you need to know how it is called. You will not directly ask an object, but the `Serial` class itself:

```
String nameSerial = Serial.list()[0];
```

The methods of this type are called *static* or *class methods*. In this case you can ask the class the name of the first serial port, which is usually the one used by Arduino.

But Not Always...

The Arduino is usually, but not always, found on the first serial port. If Processing can't find your Arduino on the first port, try changing the index to [1], [2], or something else. If you get stuck, add this line to your Processing sketch and run it (the serial port listed in the Processing console will match the one you use in the Arduino IDE):

```
println(Serial.list());
```

When you find your Arduino's port in the list, start counting from 0 until you reach the one corresponding to the Arduino. This is the number to use in place of 0 when you define `nameSerial`.

You'll use `nameSerial` to create your `Serial` object:

```
SerialPort =
new Serial(this, nameSerial, 9600);
```

As you saw in "Arduino and the Raspberry Pi" on page 182, `9600` indicates the speed at which you want the port to work. Make sure that it corresponds to the speed you have used in the Arduino sketch. You want Processing to activate itself only when Arduino has finished transmitting its data, and you can make sure that nothing happens until you press the Enter or Return key:

```
SerialPort.bufferUntil('\n');
```

In this way Processing temporarily stores all data coming onto the serial port into a *buffer*—a temporary container. When the "enter" character arrives, Processing automatically calls the se `rialEvent` method, where you can read what is written in the buffer, excluding the Enter/Return character you are not interested in:

```
dataRead = SerialPort.readStringUntil('\n');
```

Remove spaces and tabs from the string:

```
dataRead = trim(dataRead);
```

Before being able to use the read value, you have to turn it into a number, remembering that on analog ports the Arduino reads values from 0 to 1023. If you wanted a radius (the variable `ray`) between 10 and 100 you should use the `map` function, which allows you to pass from one scale of values to another:

```
ray = map(int(dataRead), 0, 255, 10, 100);
```

Example 19-6 shows the complete listing. Before you can use this, set up your Arduino with the circuit shown in Figure 16-12, and load the sketch shown in Example 16-9.

Example 19-6 *Reading from the serial port*

```
import processing.serial.*;
Serial SerialPort;
String dataRead;
float ray = 10;
void setup()
{
  size(200, 200); //make your canvas 200 x 200 pixels big
  SerialPort = new Serial(this, Serial.list()[0], 9600);
  SerialPort.bufferUntil('\n');
}
void draw()
{
  background(200);
  ellipse(width/2, height/2, ray, ray);
}
void serialEvent( Serial SerialPort)
{
  dataRead = SerialPort.readStringUntil('\n');
  if (dataRead != null)
   {
    dataRead = trim(dataRead);
    println(dataRead);
   }
  ray = map(int(dataRead), 0, 255, 10, 100);
```

```
    println(ray);
}
```

Let's conclude the chapter with some exercises:

- Modify the code of Example 19-2 to make sure that the circles bounce off not when their center reaches the border of the display, but when the perimeter touches it.
- Write a sketch so that at each click of the mouse, a line gets drawn that links the point where the pointer currently is to the last point that was created. At the first click, the sketch has to draw a point. The thickness, color, and opacity of each line (see Processing's `stroke` function) must be random. After 20 clicks the screen has to clean itself.
- Modify the last version of `DropTest` to make sure that, next to the drops created with a mouse click, every now and then a new drop appears in a random point of the display.

The Internet of Things

20

In the previous chapters we got acquainted with the tools needed to create many different things. We have used additive and subtractive technologies to build physical objects, and we have made them interactive through the use of computers and computer languages. So, what's our next step?

Physical Computing

Physical computing is concerned with the creation of interactive physical environments and systems, using hardware and software that can perceive what is happening in the surrounding area and react accordingly. People who deal with this subject first try to study the relationships between people and systems, in order to simplify and make the interactions between humans and objects as natural as possible.

Neil Gershenfeld, director of MIT's Center for Bits and Atoms (CBA), speaks about "putting the computer in your shoe"—that is, designing objects in a way that hides the complexity that accompanies technology. A physical computing project is first and foremost a design problem in which the people who are going to interact with that system are the first factor that has to be considered.

Thanks to platforms like Arduino and all the other boards that have been developed in the past few years, we can quickly create prototypes able to perceive nearly any type of external stimulus and interact with the surrounding environment by using motors, servos, network messages, or any other kind of signal.

What used to be a subject reserved for researchers and pioneers has now piqued the interest of companies, eager to explore potential applications for physical computing.

This New World

You have seen that a small computer can be embedded in an object and can interact with you and other objects. The extension of such interaction mechanisms through the Internet has given rise to the idea of the *Internet of Things* (*IoT*).

The Internet of Things opens a new world to us. For example, not only can we record our performance while we are outside jogging and transfer the data to a server for analysis and filing, but we can also take part in virtual competitions with people jogging on the other side of the globe, thus exploiting the social component as a way to involve people and objects. We can trace the journey of a package from

Asia to our front door, wondering why carriers keep taking absurd turns before finding us.

The number of connected objects was estimated in 2012 to be 8.7 billion (serving a human population of about 7 billion, remember), but the number of *connectable* things is thought to be 1.5 trillion. This disparity means that new business models and new opportunities will arise before our eyes, starting with a sort of ritual sacrifice of the nonconnected objects in favor of tools that have the same functionalities but are connected to the Net, so that we can remotely interact with them to set the temperature in our homes, turn on the oven, turn off the lights we have left on, and many other things.

We are speaking of things that are sort of aware of themselves, of their history (at least, they keep track of their state over time), and of the world surrounding them, with which they interact in a more or less integrated way.

Where to Put the Data?

Picture an environmental sensor station based on Arduino to detect various aspects of the environment. We would like to publish the data read by the sensors so that anyone can make use of them, but without having to do the publishing ourselves. Instead we'll use Xively (*https://xively.com*), which was designed for this purpose.

In this case it is really simple: we just need to register on the site and specify what kind of device we want to link. Automatically, the system generates the code for our Arduino, within which we will have to specify only where the values we want to publish come from. All we have to do is to verify the sketch and upload it —that's it!

With this service, for example, creative makers created a device (*http://marcpous.com/oktoberfest_of_things_part_ii.html*) able to trace the amount of beer they had drunk at the Oktoberfest and published the data obtained. What's more, the device is multifunctional: for example, it can also be used to calculate the number of sandwiches that accompanied the beer!

From Ivrea to Rome: Flyport

The Internet of Things is not only Arduino; there are so many other solutions. For example, the Flyport (Figure 20-1), produced by the Roman company OpenPicus, is a tiny and powerful card that allows you to connect to the Internet very easily through Wi-Fi, Ethernet, or even quadband mobile technology. We are just spoiled for choice! In this case the hardware is open source, too, and the programming environment is completely free of charge. Flyport can be extended very easily thanks to *nests*, which are similar to Arduino shields; in particular, the Grove Nest puts at your disposal many components that can be easily integrated to simplify and accelerate the prototyping of your objects.

In contrast to the Arduino IDE, OpenPicus puts a more powerful interface at your disposal: it manages all files in the project and the external libraries, provides autocompletion of code, offers the possibility to import web pages created with other tools, and includes a tool that guides you step by step in the configuration of the network parameters.

For example, in the wiki of OpenPicus (*http://bit.ly/1xPx9Tg*) you will find a very interesting hack that not only helps you determine whether you should take an umbrella with you, but also helps you not to forget it! The idea is quite simple: when you are about to leave home, a movement sensor informs Flyport, which, in turn, queries a web-based weather service in your area and, based on the answer received, colors an LED accordingly. This gives you the piece of information at the exact time and place you need it.

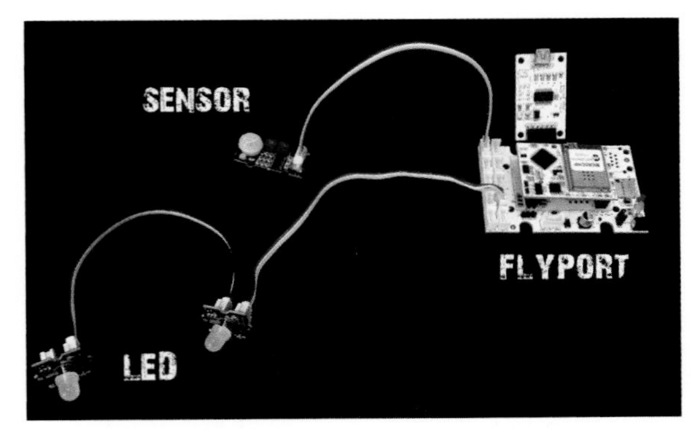

Figure 20-1 *Flyport*

Raspberry Pi on the Net

Thanks to the large number of libraries crated for it, Python allows you to perform even complex operations with a few lines of code. So why not use it in your programs to access the Internet?

The modules urllib and urllib2 provide simple methods to access online resources. With two code lines you can read the content of a web page just by using the `urlopen` and `read` methods, as shown in Example 20-1.

Example 20-1 *Python on the Net*

```
import urllib
import urllib2
service = urllib2.urlopen("http://
www.google.it/")
webpage = service.read()
print(webpage)
```

Let's try to use Raspberry Pi to query a web service. On the Internet, you can find many services you can use for free. Suppose you want to create a circuit to turn off a light at dawn. You can start with a simple LED. For this project, connect an LED to the Raspberry Pi's GPIO25 pin, just as in Chapter 18.

One of the many web services that offers all the information you need can be found at *http://www.earthtools.org/webservices.htm#sun*. Simply provide the service with some parameters in the URL address. Finding out at what time the sun rises requires the coordinates of your location (latitude and longitude), the date you are interested in, and some data on time zone and Daylight Saving Time. The parameters must be given in the following format:

```
http://www.earthtools.org/sun/<latitude>/
<longitude>/<day>/<month>/<timezone>/<dst>
```

For example, if we wanted to know at what time the sun rises in Milan on May 14th we should type the address of the service as follows:

```
http://www.earthtools.org/sun/
45.4641611/9.1903361/14/5/1/1
```

where the coordinates of Milan are 45.4641611 for latitude and 9.1903361 for longitude, 14 is the day, 5 is the month, 1 is the time zone (but you can also type 99 and the service will calculate it for us), and the last number, 1, indicates that Daylight Saving Time must be considered. Enter that URL into a web browser, and you'll get back a data dump (Figure 20-2), which you can mine for the relevant information.

earthtools.org

This XML file does not appear to have any style information associated with it. The document tree is shown below.


```
<sun xmlns:xsi="http://www.w3.org/2001/XMLSchema-instance"
 xsi:noNamespaceSchemaLocation="http://www.earthtools.org/sun.xsd">
  <version>1.0</version>
  <location>
    <latitude>45.4641611</latitude>
    <longitude>9.1903361</longitude>
  </location>
  <date>
    <day>14</day>
    <month>5</month>
    <timezone>1</timezone>
    <dst>1</dst>
  </date>
  <morning>
    <sunrise>06:08:47</sunrise>
    <twilight>
      <civil>05:36:09</civil>
      <nautical>04:55:39</nautical>
      <astronomical>04:10:18</astronomical>
    </twilight>
  </morning>
  <evening>
    <sunset>20:26:09</sunset>
    <twilight>
      <civil>20:58:56</civil>
      <nautical>21:39:41</nautical>
      <astronomical>22:25:27</astronomical>
    </twilight>
  </evening>
</sun>
```


Figure 20-2 *The XML results*

The format in which the web service provides you with the data is called *XML*, which stands for eXtensible Markup Language. It is a text format able to represent data as a complex tree structure; what's more, even though it was created to let machines communicate with one another, people can easily read it, too.

In order to read data in XML format, you need a further library; this library, though, is not simple and is composed of many subparts. Sometimes, for simplicity, you don't need to import the whole module, only the part you need. Luckily, the library's different components have been placed in a well-defined hierarchical structure, so you can identify them with names separated by dots, just like when we write the trace of a file on our hard disk. To import only part of the library, you can use the following syntax:

```
import xml.etree.ElementTree
```

Unfortunately, with this approach, each time you use `ElementTree`, you will have to repeat the complete name: `xml.etree.ElementTree`. This is really too much for a programmer! Luckily you instead use an *alias* you like, for example `etree`:

```
import xml.etree.ElementTree as etree
```

Then you will need the `datetime` object of the `datetime` module:

```
from datetime import datetime
```

With the `from ... import` syntax, in the rest of the code you can simply write:

```
datetime.strptime(...)
```

To control the LED, you will use the GPIO libraries. Then you will name the web service `urlopen`, giving the address and the parameters to use. The result will be saved in the `response` variable. The program must work every day, so you will have to name the service using today's date:

```
now = datetime.now()
service = urllib2.urlopen(
  "http://www.earthtools.org/sun/
45.4641611/9.1903361/"
  + now.day + "/" + now.month + "/1/1")
response = service.read()
```

The `fromstring` function of the `xml` module builds a data tree in memory, starting from the result of the call to the service that we insert in the tree variable:

```
responsetree = etree.fromstring(response)
```

You need to get the information on the sunrise time. The tree built by Python has a main node (or *root*), <sun>, under which there are five more nodes: <version>, <location>, <date>, <morning>, and <evening>, as shown in Figure 20-3.

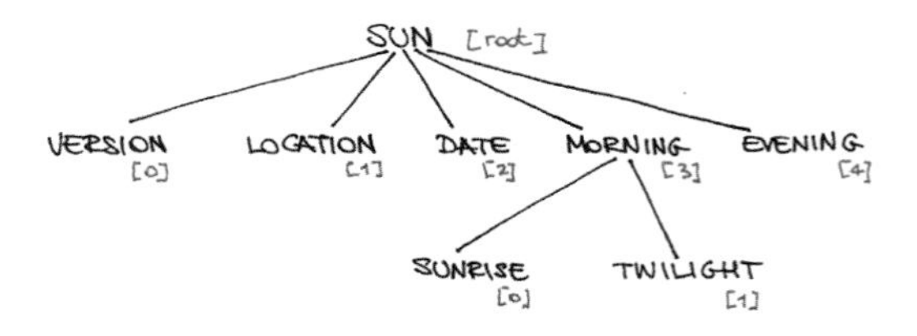

Figure 20-3 *The data tree provided by the service*

The information you need is in the <sunrise> node, placed under the <morning> node, the fourth one. Because Python counts starting from 0, we use an index of 3.

By writing `tree[3]` you will place yourself in the <morning> node, which includes the <sunrise> and <twilight> nodes. The <sunrise> node is the first one, so the index to use is 0:

```
responsetree[3][0]
```

The time you are interested in is included between the `<sunrise>` and `</sunrise>` *tags*. To extract the text between the two delimiters we will use the `text` method:

```
responsetree[3][0].text
```

Now you have some text with the time of sunrise. Unfortunately, for a computer this text is not a date or a time: you have to convert it. To do so you will use the `datetime` module. The function to make conversions doesn't have a very intuitive name: `strptime`:

```
datetime.strptime(
  responsetree[3][0].text, '%H:%M:%S')
```

The `strptime` function needs two parameters: the text to convert and its format. In the

06:08:47 string the first two digits express the hour, followed by a separator (:) for the minutes and a further separator for the seconds. Tell strptime that the time is expressed in hours with %H, while %M indicates minutes and %S seconds. Now the function is able to understand and create a datetime object that we will call wakeup. But you are still only halfway there…

A datetime object is a sort of a watch: it memorizes a date and a time. When you created wake up, you set its time but not its date: let's copy today's date, which is the same date you used when we called the service. To copy year, month, and day in the wakeup object, use the replace method:

```
wakeup = wakeup.replace(now.year,
  now.month,
  now.day)
```

Now it does make sense to compare wakeup and now! If the date is the same and it is before the sunrise, the light must be on; otherwise, it must be off. See Example 20-2 for the full code.

Example 20-2 *When does the sun rise?*

```
import urllib
import urllib2
import xml.etree.ElementTree as etree
from datetime import datetime
import RPi.GPIO as GPIO

GPIO.setmode(GPIO.BCM)

#the GPIO25 pin will be used as output
GPIO.setup(25, GPIO.OUT)

#creates the datetime object corresponding to the present
now = datetime.now()

service = urllib2.urlopen(
  "http://www.earthtools.org/sun/45.4641611/9.1903361/" + str(now.day)
  + "/" + str(now.month) + "/1/1")
response = service.read()

#parses the xml from a string:
responsetree = etree.fromstring(response)
print "The sun rises at: " + responsetree[3][0].text

#converts the time from text into a datetime object
wakeup = datetime.strptime(responsetree[3][0].text, '%H:%M:%S')

#wakeup hasn't set the date: we need to copy it from now
wakeup = wakeup.replace(now.year, now.month, now.day)
if (now > wakeup):
    print "turn off the light"
    GPIO.output(25, GPIO.LOW)
else:
    print "turn on the light"
    GPIO.output(25, GPIO.HIGH)
```

Things to try: modify the script so that the light turns on after the sunset. The information on time can be found somewhere in the twilight node.

Features of a Service

What have you learned about creating an Internet of Things service? It all boils down to eight elements. Some IoT projects will have all eight, while some will focus only on a few. How many of these elements does your IoT project have?

Events
: The starting point is an event we are interested in. It can be pressing a button, the temperature rising or falling, or the noise level exceeding a specific limit. It can also be the passing of time, as with the smart containers of medicines that remind patients when it is time to take a pill.

Controllers
: A controller oversees the environment in which the event takes place. The controller is equipped with tools able to recognize the occurrence of the event, and knows how to behave accordingly. In general, controllers are the objects interacting with us and with the system as a whole.

Sensors
: To monitor events, controllers use a series of sensors, each of which is specialized to detect a particular kind of event: thermistors for temperature, photoresistors for light, proximity sensors for distance, even down to simple on/off buttons. Whatever it's measuring, the sensor only generates the signal; it relies on the controller to process the signal correctly.

Log
: IoT objects are (or should be) aware of their own history. When a sensor generates a signal, the controller doesn't just communicate that signal; it can also record the event on a storage medium somewhere. In this way, it may be possible, for example, to repeat a series of events as you wish, as you could do with *motion capture*.

Upload
: The Internet of Things is nothing if the data can't reach the Internet. Controllers collect data from their sensors and then make them available to the rest of the world via the Internet.

Analysis
: Controllers usually don't have the power to analyze data. That analysis, if it happens at all, is performed on another computer after the data has been uploaded. After the analysis you can decide which behavior the system must adopt.

Network effects
: The social component has a strong influence on defining success or failure within the Internet of Things. Data become not only information, but also occasions for meetings and exchanges that will end up enriching complex systems, thus increasing their value over time.

Action
: If the analysis dictates that a certain action must be carried out, the controller makes sure that the corresponding actuator operates accordingly: it may be something simple, like a servomotor rotating based on how many tweets with a specific hashtag appear in a specific timeframe, or complex actions involving many other systems.

The power of the Internet of Things lies is its system of loosely coupled components—that is, those that are easily interchangeable with other components with the same interface, and so exhibit the same behavior.

In this context the most important point you must keep in mind is always the user's benefit: in the Internet of Things everything is a service. We must never forget that! To quote Chris Anderson, "We are all designers now, we might as well get good at it."

Index

Symbols

123D Design software, 78, 93, 110
3-axis CNC machines, 100, 102
3D Hubs i.materialise, 98
3D printing
  basic printer operation, 89
  benefits of, 90
  detailed printer operation, 90
  economic impact of, 13
  increased access to, 13, 89
  material selection, 90
  object creation with, 77
  online printer reviews, 93
  online sources for, 98
  printer selection, 91
  technologies available, 89
  workflow
    considering tolerances, 93
    correcting defects, 93
    G-code translations, 95
    model creation, 93
    object finishing, 98
    printer operation, 96
    printer set-up, 95
3D scanners, 93
3DTin design software, 79
= (equals sign), 149
== (double equals sign), 149

A

acrylonitrile-butadiene-styrene (ABS), 90
actuators
  buzzers, 161
  purpose of, 161
  servo motors, 162
Adafruit, 129
additive manufacturing, 77
aesthetics, 32
affordances, 33
Alexander, Christopher, 33
ampere (A), 125
angel investors, 56
animation, 188
Arduino
  applications for, 141
  benefits of, 142, 201
  development of, 141
  IDE for, 142
  shields for, 164
  sketches
    adding conditions, 148
    adding contingencies, 146
    adding pauses, 146
    adding repetitions, 149
    basic syntax, 142, 144
    control structure, 148
    data storage with variables, 147
    defining pin behavior, 145
    for analog sensors, 152
    LED blink program, 143
    practice exercises, 158
    premade examples, 143
    pulse-width modulation, 156
    serial monitor, 155
    switching on/off, 146
    uploading, 144
    verifying, 144
  software structure, 142
  using with Processing, 198
  using with Raspberry Pi, 182
  vs. Raspberry Pi, 167
  wearable textiles and, 164
  working with actuators, 161-163
  working with sensors, 159-161
  working with strong currents, 163
ARPANET project, 12
arrays, 190
art, generative, 78
artifacts, definition of, 23
artisans, 59
assets, 46
associations, creative thought through, 19
assumptions, changing, 20
Autodesk 123D Catch, 93

Autodesk 123D Make, 110
Autodesk Fusion 360, 103

B

BCM numbering, 180
biological factors of design, 32
bitmap images, 112
Blender design software, 78
bootstrapping, 57
boundaries, shifting, 22
Bowyer, Adrian, 91
branches, 73
breadboards, 136
bridging, 97
buffers, 199
build-measure-learn cycle, 49
business model canvas, 51
business plans
  abstract section, 45
  assets section, 46
  business model canvas, 51
  financial section, 46
  for startups, 47
  management/organization section, 46
  marketing plan section, 46
  operating plan section, 46
  pitfalls of traditional, 46
  product/service section, 45
  purpose of, 45
buttons, 134
buzzers, 161

C

CAD software, 103
CAM software, 103
capacitors, 133
carving, types of CNC, 102
cathodes, 130
circuits
  creating a basic, 124
  creating with breadboards, 135
  creating with matrix boards, 137
  hydraulic model , 128
  LED blink program, 123
  measuring current and voltage, 137
  power source, 124, 137
class methods, 198
classes, 193
CNC machines
  applications for, 100
  benefits and drawbacks of, 107
  building your own, 108
  CAD software for, 103
  CAM software for, 103
  control software for, 107
  designing with, 102
  desktop versions, 99, 107
  drivers and encoders in, 101
  materials milled, 100
  milling heads, 100
  outsourcing, 108
  subtractive manufacturing with, 77, 99
  variations of, 100
collaboration
  benefits of distributed intelligence, 60
  challenging aspects of, 60
  craftsperson guilds, 59
  Creative Commons licenses, 61
  of digital information, 62 (see also project files)
  role of Internet in, 12, 59
collective intelligence, 59
color, vs. colour, 190
committed files, 68
company structure, 46
components
  basic tools for, 135
  buttons and switches, 134
  capacitors, 133
  datasheets, 128
  diodes, 134
  LEDs, 130
  obtaining, 129
  overview of, 129
  part numbers/markings, 128
  resistors, 131
  symbols and diagrams for, 130
  trimmers/potentiometers, 132
Computer Aided Design (CAD), 78
computer numerical control (CNC) (see CNC machines)
concept selection matrix, 30
conductors, 125
control structure, 148
craftspeople, 59
Creative Commons licenses, 61
creative techniques
  experimentation, 19
  generating alternative solutions, 19
  lateral vs. vertical thinking, 19
  making associations, 19
  mind maps, 20
  networking, 19
  overturning assumptions , 20
  pseudorandom input, 22
  role of learning in, 18
  role of neurophysiology in, 17
  shifting boundaries, 22
  sources of creativity, 17
crowdfunding, 56
Cupcake 3D printer, 92
current, 125, 137, 163
CVS, 66

D

DARPA (Defense Advanced Research Projects Agency), 12
datasheets, 128
debugging, in Python, 176
decomposition approach, 28
design
  aesthetics in, 32
  definition of, 23
  design patterns, 33
  distributed, 65
  for laser cutting, 119
  form and function in, 29
  objective of, 23
  software for, 78
  with CNC machines, 102
die-cutting machines, 77
DigiKey, 129
digital models
  correcting defects in, 93
  creating, 78

obtaining, 93
diodes, 134
direct observation, 27
distributed design, 65
distributed intelligence, 60
distributed version control, 66
divide et impera (divide and conquer), 39
Do It Yourself (DIY) technology, 7
dominant ideas, 26
double equals sign (==), 149
Dougherty, Dale, 9
drawtext command, 87
drivers, 90

E

e-textiles, 164
EBITDA plan, 46
Economic Development Corporations, 56
electroluminescence, 130
electromechanical calculators, 12
electronics
  alternative view of, 123
  circuits, 128, 135
  components, 128-135
  current, 125
  electricity basics, 125
  Hello World application, 123
  measuring electricity, 137
  Ohm's law, 139
  resistance, 127
  safety issues, 127
  voltage, 126
  voltage dividers, 133
electrons, 125
elegance, 30
Element 14, 129
EMC2 LinuxCNC, 107
endmill tool, 100
ENIAC computer, 12
Enigma machine, 12
Eno, Brian, 22
enterprise philosophy, ix
entrepreneurs, new approach of, ix
equals sign (=), 149
evaluating alternatives, 29-32
events, 192
experimentation, creative thought through, 19
extruders, 90, 96

F

Fab Labs, 7, 8, 110
Farnell, 129
file management (see project file management)
financial plans, 46
five whys method, 25
Flyport, 202
focus groups, 27
for loops, 150
for statements, 187
forgery, 62
form and function, 29
Frankenstein Garage, x
FreeMill CAM software, 103
freemium services, 66
Fritzing, 130
funding sources
  angel investors, 56
  bank loans, 55
  bootstrapping, 57
  crowdfunding, 56
  Economic Development Corporations, 56
  friends and family, 55
  selecting, 57
  venture capital, 56
fused deposition modeling (FDM), 89

G

G-code, 95, 106
Gantt charts, 41-43
Gantter software, 42
GanttProject software, 42
genchi gembutsu principle, 50
generating alternatives, creative thought through, 19
generative art, 78
Gershenfeld, Neil, 8
getting help, xi
Git
  basic workflow in, 69
  branch feature, 73
  detailed workflow in, 70
  installing locally, 69
  project creation, 66
  purpose of, 66
  three stages of work in, 68
Good-Fast-Cheap triangle, 38
Grasshopper plug-in, 78
guilds, 59

H

hackerspaces, 7, 110
Hello World program, 123, 176, 180, 185
hypothesis of growth, 49
hypothesis of value, 49

I

i.materialise prototyping service, 98
Industrial Revolution
  effect of computers on, 12
  short history of, 11
infill, 97
information hiding, 197
information, increased access to, 12
  (see also collaboration)
Inkscape CAD software
  creating a circle in, 113
  creating projects in, 113
  downloading, 112
  for 3D printing, 103
  optimizing files, 118
  tricks for 3D, 119
  tutorials for, 112
  working with images in, 116
  working with text in, 113
innovation
  importance of, 5
  techniques for, 18
innovative projects, 35
input, pseudorandom, 22

instances, 193
Instructables, 59
insulators, 125
intellectual property, 61
Internet
  impact on personal
    manufacturing, 13
  importance to collaboration, 59
  origins of, 12
Internet of Things (IoT)
  concept of, 201
  features of a service, 206
  networking connections, 202
  physical computing, 201
  publishing data, 202
  Raspberry Pi and, 203
  strength of, 206
iron triangle, 38
iteration
  in design process, 34
  in project management, 39

J

Jameco, 129
jumpers, 136

K

Kentstrapper 3D printer, 92
kerf, 118
Kirchhoff's current law, 128

L

laser cutting
  3D objects and, 119
  applications for, 110
  community resources for, 110
  iterative approach to, 119
  model creation for, 112-119
  online sources for, 110
  possible materials, 109
  reducing costs of, 118
  software for, 110
  tutorials for, 119
  types of cutters, 109
Lasersaur laser cutter, 110
latent needs, 27
lateral thinking, 19
Lean Startup movement, 49
learning process, role in creativity,
  18
LED blink program
  Arduino controlled, 143
  basic circuit for, 124
  capacitors and, 134
  materials required, 123
  vs. Hello World , 123
LEDs, 130, 165
legal protection, 61
licenses, 61
life projects, 37
light amplification by stimulated
  emission of radiation (see laser
  cutting)
light sensors, 160
LilyPad Arduino , 165
logical numbering, 180
loops, 150

M

MACH3 for Windows, 107
MailChimp, 57
MAKE Magazine, 9
Maker Faires, 9
maker movement
  contributions of Neil
    Gershenfeld to, 8
  culture of reuse in, 3
  culture of sharing in, 8, 59
  definition of, ix, 5
  economic impact of, 5, 11-14
  effect of technology on, 4, 8
  MAKE Magazine, 9
  origins of, 7-9
  skills required for, ix, 4
  social aspects of, 9
MakerBot 3D printers, 92
MakerShed, 129
makerspaces, 7, 60, 110
management plans, 46
  (see also business plans)
manufacturing process
  3D printing, 77, 89-98
  additive vs. subtractive, 77
  digital model creation, 78, 93
  laser cutting, 109-119
  milling machines, 99-108
  OpenSCAD software, 79-87
marketing plans, 46
matrix boards, 137
matrix of concept selection, 30
Mebotics Microfactory, 100
MechSoft FreeMill, 103
memory, nonvolatile, 142
Mendel Max 3D printer, 92
Mercurial, 66
MeshLab, 95
MeshMixer, 95
Microsoft Kinect, 93
Microsoft Project, 42
milestones, 42
milling machines
  benefits and drawbacks of, 107
  CNC, 99
  endmill tool, 100
  manufacturing using, 77, 99
  milling heads, 100
  software for, 103
mind maps, 20
minimum viable product (MVP),
  49
Mintronics: Survival Pack, 124
modified files, 68
modularization, 100
MOSFET (metal-oxide-
  semiconductor field-effect
  transistor), 163
Mouser, 129
multimeters, 137

N

nano text editor, 176, 181
needs, latent, 27
nesting, 118
nests, 202
Netfabb Studio Basic, 94
networking, creative thought
  through, 19

neurophysiology, role in creativity, 17
nonvolatile memory, 142

O

object-oriented programming (OOP)
 borrowing classes, 197
 classes and objects, 193
 concept of, 192
 creating classes, 195
 creating multiple objects, 196
 creating objects, 194
 modifying classes, 197
objects, 193
observation, 27
Ohm's law, 127, 139
open source phenomenon
 benefits of, 5
 challenging aspects of, 60
 culture of sharing in, 8, 59
 effect of, ix
OpenPicus, 202
OpenSCAD
 basic interface, 80
 benefits of, 79
 combining objects in, 85
 creating 3D objects in, 86
 creating cubes in, 80
 creating ellipses in, 86
 creating spheres in, 80
 exporting models for printing, 80
 modifying resolution in, 81
 removing objects in, 85
 repeating commands, 82
 rotate function, 84, 86
 scaling object in, 84
 storing information in, 81
 text libraries for, 87
 translate function in, 82
operating plans, 46
overhang, 97

P

parent-child projects, 9
patents, 61
patterns, in project design, 33
perpendiculars to the plane, 93
personal manufacturing, 13, 89, 92
photoresistors, 160
physical computing, 201
physical numbering, 180
piezoelectric transducers, 161
pivot/persevere phase, 50
pixels, 112
Planner software, 42
Pleasant3D, 94
polylactic acid (PLA), 91
polyvinyl alcohol (PVA), 91
Ponoko prototyping service, 98, 119
postmortems, in project lifecycle, 39
potentiometers, 132
PowerWASP 3D printer, 92
predictable projects, 35
printed circuit boards (PCBs), 137
Printrbot 3D printers, 92
Printrun, 96
problem solving, 23, 60
Processing
 adding animation with, 188
 animating multiple objects, 190
 comprehensive manual for, 187
 creating a circle with, 188
 creating a line with, 186
 downloading, 185
 Hello World program in, 185
 object-oriented programming
  basic concept of, 192
  borrowing classes, 197
  classes and objects, 193
  creating classes, 195
  creating multiple objects, 196
  creating objects, 194
  modifying classes, 197
 popularity of, 185
 practice exercises, 200
 repetitive commands in, 187
 responding to events, 192
 sketches in, 185
 using with Arduino, 198
 working with colors in, 190
 working with text in, 186
product adoption curves, 47
product development model, 47
project design
 aesthetics in, 32
 challenging aspects of, 23
 decomposition approach, 28
 definition of design, 23
 design patterns, 33
 evaluating alternatives, 29
 for laser cutting, 119
 iteration in, 34
 listing requirements, 27
 modularization, 100
 predictable vs. innovative projects, 35
 problem definition, 25
 process overview, 24
 prototyping, 33
 software for, 78
 with CNC machines, 102
project file management
 distributed version control, 66
 Git and GitHub, 66-70
 Git's branch feature, 73
 version control system for, 65
 workflow for, 70
project management
 basics of, 38
 common project attributes, 37
 common project limitations, 38
 Gantt charts, 41
 importance of, 37
 iron triangle of, 38
 listing activities, 39
 project lifecycle, 39
 risk management, 42
 software for, 42
project review meetings (PRMs), 42
prototyping, 33, 98, 142, 201
pseudocode, 148, 148
pseudorandom input, 22
pulse-width modulation (PWM), 156
Python
 accessing the Internet with, 203
 Arduino and, 183
 benefits of, 176

calculator operation, 176
code syntax, 177
connecting button switches in, 182
creating comments in, 177
creating scripts in, 176
development environment, 176
exiting, 176
GPIO library
configuring ports with, 180
controlling LEDs with, 180
repeating instructions in, 180
Hello World program in, 176, 180
interpreted language of, 176
launching, 176
official documentation for, 177
Raspberry Pi and, 167
RPi library
logical vs. physical numbering in, 180
managing GPIO pins with, 179
time library in, 180
variables in, 177
vs. other languages, 176

R

raft layer, 98
rails, 136
Raspberry Pi
basic shell commands, 171
benefits of, 167
component overview, 167
files and directories in, 172
GPIO pins, 177-182
graphical user interface, 175
initial set up, 169
Internet access, 203
monitoring peripherals, 174
Python interpreter, 176
redirection, 173
superusers, 174
using with Arduino, 182
vs. Arduino, 167
raster images, 112
Redmine software, 42
relief grooves, 119
Repetier-Host, 96
replicating rapid prototyper, 91
Replicator 3D printer, 92
repositories, 66
RepRap printer, 91
requirements, determining, 27
resistance, 127
resistors, 123, 131
restrictions, in project design, 25
Rhinoceros design software, 78
risk management, 42
Rory's Story Cubes, 22
RPi library, 179

S

safety issues
importance of training, 8
working with electricity, 127
scanners, 93
Sculpteo prototyping service, 98
sensors
available types, 160
light, 160
photoresistors, 160
purpose of, 159
temperature, 159
thermistors, 159
serial monitor, 155
servo motors, 162
Shapeways prototyping service, 98
shells, 97
shields, 164
Simple Maker's Kit 3D printer, 92
sketches, 142, 185
SketchUp Make design software, 78
Slater Technology Fund, 56
slicing, 95
smart textiles, 164
snap-fit joints, 100, 119
solutions
evaluating alternative, 29
finding alternative, 19
SparkFun, 129
sponsored funding, 57
staged files, 68
stakeholders, 42
Standard Tesselation Language (STL), 78, 80
static methods, 198
stepper motors, 90
Structure Scanner, 93
subtractive manufacturing, 77
Subversion, 66
SVG (Scalable Vector Graphics), 112
switches, 134
system variables, 187

T

tacit knowledge, 30
technical approach, 33
technology
benefits of, 4, 8
DIY renaissance, 7
effect on Industrial Revolution, 12
increased access to, 4, 7, 13
limitations of, 3
TechShop, 7
temperature sensors, 159
TextGenerator.scad file, 87
thermistors, 159
Thingiverse, 13, 61, 93
time library, 180
TinkerCAD design software, 79
tinkerers, 4
tolerances, 93
trademarks, 62
trimmers, 132

U

Ultimaker 3D printer, 92
uncertainty cone, 43

V

value propositions, 52
variable resistors, 132

variables, 147
vector graphics, 112
(see also Inkscape CAD software)
venture capital, 56
version control system (VCS), 65
vertical thinking, 19
visual representations (see Processing)
volt (V), 126
voltage, 126, 137
voltage dividers, 133

W

WASP (World's Advanced Saving Project), 92
watertight models, 94
wearable textiles, 164
Weebly, 57
WordPress, 57
work breakdown structure (WBS), 39
wow factor, 30

X

Xively, 202
XML (eXtensible Markup Language), 203

Y

YouMagine, 13, 61, 93

About the Authors

Andrea Maietta is a passionate advocate of agile methods, and is responsible for helping clients understand their needs and providing them with appropriate solutions to build value. He is a software engineer, maker, tireless reader, husband and father, and Rugby Union fan for life. Always excited to learn and share, Andrea is a frequent speaker at conferences and is involved in training, communication, and organization. His dream is to build a lightsaber.

Paolo Aliverti became interested in electronics and microcomputers at the age of ten. He is one of the founders of Frankenstein Garage, where he works on design and research as well as agile business, marketing, and planning. He organizes workshops on electronics and 3D printing, participates in conferences and events, and explains in simple words how easy it is to rebuild and repair items. In the little free time he has, he risks his life in the mountains.

Colophon

The cover and heading font is Benton Sans, the body font is Myriad Pro, and the code font is Bitstream Vera Sans Mono.